Lecture Notes in Networks and Systems 1272

The series "Lecture Notes in Networks and Systems" publishes the latest developments in Networks and Systems—quickly, informally and with high quality. Original research reported in proceedings and post-proceedings represents the core of LNNS.

Volumes published in LNNS embrace all aspects and subfields of, as well as new challenges in, Networks and Systems.

The series contains proceedings and edited volumes in systems and networks, spanning the areas of Cyber-Physical Systems, Autonomous Systems, Sensor Networks, Control Systems, Energy Systems, Automotive Systems, Biological Systems, Vehicular Networking and Connected Vehicles, Aerospace Systems, Automation, Manufacturing, Smart Grids, Nonlinear Systems, Power Systems, Robotics, Social Systems, Economic Systems and other. Of particular value to both the contributors and the readership are the short publication timeframe and the world-wide distribution and exposure which enable both a wide and rapid dissemination of research output.

The series covers the theory, applications, and perspectives on the state of the art and future developments relevant to systems and networks, decision making, control, complex processes and related areas, as embedded in the fields of interdisciplinary and applied sciences, engineering, computer science, physics, economics, social, and life sciences, as well as the paradigms and methodologies behind them.

Indexed by SCOPUS, EI Compendex, INSPEC, WTI Frankfurt eG, zbMATH, SCImago.

All books published in the series are submitted for consideration in Web of Science.

For proposals from Asia please contact Aninda Bose (aninda.bose@springer.com).

Francesca Moraci · Carmelina Bevilacqua ·
Pasquale Pizzimenti
Editors

Ecological and Digital Transition in Cities

Measuring Ecosystem Services for Urban Planning and Design

Springer

Editors
Francesca Moraci
DArTe Department
Università degli Studi Mediterranea di Reggio Calabria
Reggio Calabria, Italy

Carmelina Bevilacqua
PDTA Department
Sapienza Università di Roma
Rome, Italy

Pasquale Pizzimenti
DArTe Department
Università degli Studi Mediterranea di Reggio Calabria
Reggio Calabria, Italy

ISSN 2367-3370 ISSN 2367-3389 (electronic)
Lecture Notes in Networks and Systems
ISBN 978-3-031-82926-0 ISBN 978-3-031-82927-7 (eBook)
https://doi.org/10.1007/978-3-031-82927-7

This Springer imprint is published by the registered company Springer Nature Switzerland AG
The registered company address is: Gewerbestrasse 11, 6330 Cham, Switzerland

Preface

Planning the Transition for Cities and Territories: Innovative Approaches and Research Trajectories

The high level of uncertainties and complexity under the continuous shocks and stresses that have arisen in the last years have led the European Union to set out a strategic agenda (2024–2029) grounded on three main pillars [1]: a free and democratic Europe, a strong and secure Europe, a prosperous and competitive Europe. Such strategic pillars influence the EU's complex set of policies built upon an integrated policy approach that simultaneously addresses environmental (biodiversity, climate change, energy), economic (technology and digitalization, industrial transformative development, strategic investments, circular economy), and social (cohesion) dimensions in the context of climate neutrality and resilience-oriented development. The environmental dimension has assumed centrality in the complex set of EU policies. Natural capital protection and restoration, biodiversity enhancement, and energy and climate change issues are at the core of the economic restructuring processes [2]. They are pursued under the overall main objective of reinforcing social, economic, and territorial cohesion. The recent Nature Restauration Law [3] is a central element of the EU Biodiversity Strategy, whose main aim is to restore natural resources to increase biodiversity and secure ecosystem services (ESs), contributing to limiting climate change and increasing Europe's resilience. These elements are also included in the European Green Deal, the ambitious plan to transform Europe into the first climate-neutral continent. The Green Deal is a strategic framework to accelerate the transition toward a carbon-neutral economy by decoupling economic growth from resource exploitation. It sets ambitious objectives by 2050, promoting sustainability and resilience in several thematic areas through an integrated approach [2]. Activating such a transformative shift generates a new demand for sustainability in cities—which catalyzes restructuring processes—for balancing economic development and natural capital enhancement crucial for human well-being. The pressures generated on natural ecosystems are harnessing their ability to produce the ESs vital for humans [4–6], and urbanization processes have simultaneously increased the demand for natural capital in cities [7]. From regulating climate and air quality to enhancing biodiversity and recreational spaces, ESs are nowaday integral to urban health and sustainability [8, 9]. However, integrating these services into the urban fabric requires innovative thinking and a departure from conventional planning paradigms [10]. ESs include providing services like food and water, regulating services such as climate regulation and flood control, cultural services providing recreational, aesthetic, and spiritual benefits, and support services, including nutrient cycling and soil formation [10]. Also, in the urban context, implementing ESs can mitigate urban heat islands, improve air and water quality, and enhance the overall well-being of city dwellers.

The main challenge lies in effectively integrating these services into urban planning and policy for their reinforcement by considering the complex interdependencies between human and natural systems [11, 12]. In this direction, this book advocates for a paradigm shift in planning the transition of cities and territories, focusing on the centrality of natural capital for their future. A critical aspect of such integration lies in digital technologies and data-driven approaches. The advent of big data, urban informatics, and advanced modeling techniques (artificial intelligence and machine learning) offers unprecedented opportunities in this direction. By leveraging these technologies, planners and policymakers can make informed decisions that optimize the delivery and benefits of ESs [13]. This data-driven approach enhances the precision and efficiency of urban planning and facilitates the monitoring and evaluation of ecosystem service outcomes over time.

These elements pose a fundamental question: What kind of planning is required during times of transition? How can it strengthen the sustainable and resilient development of cities and regions amidst persistent shocks and pressures? How can it revolutionize its methodologies and champion innovative approaches to address such complex challenges?

Urban planning encounters an impasse phase [14]. Starting from late 2000, with the financial crisis of 2008, the role of planners and the ability to plan to preconfigure sustainable futures dealt with the neoliberal agenda [15], which has shown during and after the pandemic that inequalities, democratic gaps, and social exclusion increased. Rapid urban growth in recent decades has led to a dichotomy between competitive development and social and environmental issues [15]. Economic development, driven by neoliberal approaches, has misaligned welfare principles in urban planning, emphasizing more market-oriented dynamics [16, 17]. This has increased social, economic, and environmental vulnerabilities, and traditional urban transformation practices aimed at boosting competitiveness have become inadequate, especially in the face of climate change, emphasizing the pressing request for equity in transitioning. Since the 1990s, urban regeneration has emerged as a crucial element in urban planning, integrating social, economic, and environmental aspects to enhance city sustainability and resilience [18]. The evolution of urban regeneration and approaches has contributed to developing new forms of public-private partnerships, citizen-administration interactions, and mixed-use developments. These strategies aims to promote economic and social inclusion and catalyzed environmental sustainability. Urban regeneration shows the potential to address global challenges and support the local ecologicla, digital, and inclusive transitions [19]. Such characteristics open innovative perspectives on planning cities' transition in stressing adaptive and regenerative approaches in urban planning: two forward-thinking approaches that aim to respond to current challenges and anticipate future needs beyond sustainability and resilience.

Adaptive approaches focus on flexibility, enabling cities to respond to changes and uncertainties, emphasizing the ability of urban systems to quickly adapt to new conditions and continue to function effectively [20, 21]. Urban plans in this context involve urban regeneration initiatives that aim to create strategies and interventions to withstand and recover from various shocks and stresses. The goal is to increase urban resilience and address critical vulnerabilities. By using data-driven approaches to construct new scenarios, effective, flexible urban strategies can be developed to adapt to different conditions. This requires dynamic and iterative processes to ensure that urban plans remain relevant and effective over time. Engaging stakeholders, including citizens, businesses, and policymakers, is also crucial to ensure cohesion and support for adaptive measures.

Regenerative approaches go beyond sustainability, including it in a systemic, holistic, and integrated approach aimed at bouncing back development conditions under an acceptable threshold represented by the earth's capacity to fulfill human needs [22]. At the core of these approaches lies the idea of creating the conditions for a different and more balanced nature-human interaction centered on circular metabolism [23]. Applying such concepts in urban planning processes and tools is challenging. The urban population is constantly growing and will continue to grow in the next decades, demanding natural capital [7] but also for housing, energy, transportation, increased economic activities, and urban infrastructures. Planning and Policies instruments have the difficult task of accommodating such needs and, at the same time, deploying suitable solutions for sustainable development. Concepts such as "density," "compact development," and "mixed-use" are being revisited in an attempt to optimize urban environments and reverse the negative trend activated by economic-growth-oriented models.

The adaptive and regenerative approaches outline a clear tendency to revise and innovate planning practices for cities and territories in response to the demand for resilience (social, economic, and environmental) and sustainability (which embrace a new dimension under the regenerative conceptualization). Both approaches are somehow interrelated and share the common root of preserving and restoring natural capital. In this direction, understanding the complex interaction between nature and human activities characterized by the social, built, and natural capital components provides the ground for upgrading planning approaches and practices. In such interactions, ESs play a central role in human well-being, and can drive policies, planning, and planning instruments in the short, medium, and long term.

The pace and intensity of the pressures generated by climate change call for urgent mitigation and adaptation measures, and cities can contribute to finding solutions [24]. They host most of the world's population, concentrate most of the emissions of pollutants, and are the places of socio-economic inequalities [4]. At the same time, intervening in cities offers the chance to address such complex issues and generate positive "spill-over" effects in the urban-rural relationship by rebalancing the pressures of urbanized areas on the surrounding context through the deployment of innovative solutions [4, 5].

Therefore, the book explores the potential of data-driven methodologies for developing innovative urban planning practices and approaches for resilient, sustainable, and equitable cities facing the challenge of planning their ecological, digital, and inclusive transition. In the face of unprecedented urban growth and urbanization side effects, environmental challenges, and socio-economic disparities, cities worldwide are at a critical point. The call for an integrated approach that combines ecological sustainability with economic development by ensuring social inclusion and equity has never been more pressing. The book aims to engage the discussion on how to shape the ecological, digital, and inclusive transition of cities and territories towards sustainability and resilience, starting from the centrality of urban planning in the promotion of such a transformative development. Primarily, it targets the academic and policymakers communities that are navigating the transition's complexity under the challenges arising after the pandemic, such as geopolitical events, political instability, energetic issues, and democratic and representative challenges, which are questioning the essence of the economic development paradigm pursued so far and calling for reshaping development trajectories through innovative planning approaches.

The book reflects the ongoing synergistic activities of three Next Generation EU-funded projects under the Italian National Recovery and Resilience Plan (NRRP). The Pilot Project 4.6.1 of the Tech4you Innovation Ecosystems (Mission 4 - Component 2 - Investment 1.5), the ECO-SET project - A Multidisciplinary approach to plan ECOsystem SErvices for cities in Transition (Mission 4 - Component 2 - Investment 1.1), and the PLANET – Planning ecosystem services for cities in transition (Mission 4 - Component 2 - Investment 1.2). All these projects collectively aim to explore and better understand how to promote a data-driven approach in urban planning and design stemming from the relevance of natural capital and biodiversity in facing the challenges of climate change for the transition of cities and territories. The Pilot Project 4.6.1 of the Tech4You Innovation Ecosystem focuses on innovative solutions for addressing urban and territorial fragmentation, particularly in southern Italian regions (Basilicata and Calabria). The project aims to implement dynamic, site-specific interventions that cater to the evolving demand for sustainable and effective transformations by promoting green and blue infrastructures. These interventions also address the challenges of depopulation and decentralization, enhancing the urban-rural connection by developing AI-based predictive models and scenarios for developing climate-resilient urban planning. The ECO-SET project focuses on developing a data-driven approach to urban planning to enhance ESs within urban transformations. Based on the Natural Capital Approach, according to the principles of biodiversity, ESs allow for improving the supply of goods and services for the well-being of society by incorporating the ability to adapt to both current risks and future climate change, reducing the ecological footprint and ecological debts, while improving resilience, health and quality of life. The scope is to operate a technological nexus between Territorial Intelligence (TI) and zoning rules to frame planning models for handling the complex systems involved in climate-proofing toward a user-tailored perspective, in which urban regeneration plays a central role for future development policies. The ECO-SET project adopts an interdisciplinary approach to experiment with the potential of data-driven urban planning in supporting the innovations necessary to make ESs a routine part of urban and infrastructure development by municipalities. The PLANET project delves into the mechanisms that trigger urban regeneration nurtured by socio-ecological-technological dynamics. This project highlights the need for alignment between digital and ecological transitions, which—by nature and characteristics—inherently follow different dynamics and time trajectories. A new perspective on urban regeneration is emerging, focusing on reducing pressures on natural ecosystems and characterized by strategic drivers such as ESs and key enabling technologies (KETs) for facilitating the transition of cities and more equitable development. Together, these projects form a collaborative effort shaping promising research trajectories in the urban planning field, providing interesting insights for evolving and innovating urban planning practices.

The proposed comprehensive integration of knowledge and methodologies is investigated to address the multifaceted challenges of urban and territorial transitions from a planning perspective. This research framework clearly shows the relevance and importance of the topics under investigation for the future of cities and regions and their sustainability and resilience. This transition pathway pivots around the three main dimensions of sustainability, namely the social, economic, and environmental, and the three main components of human life on this planet, namely the natural, social, and built capital. Dimensions and capital are the two key elements that planning has to consider for ensuring stable conditions for the current and future generations.

The abovementioned research trajectories are aligned with the evolution of the policy context in the European Union, characterized by the aim to facilitate the recovery and increase resilience after the pandemic. A policy effort finds operativeness in the National Recovery and Resilience Plans (NRRP), whose aim is to facilitate the ecological, digital, and inclusive transition of the Union. At the core of this path, the Next Generation EU instrument was introduced for recovery after the pandemic, combined with the ordinary resources of the programming framework 2021–2027 to boost investments for the transition in EU countries. However, the instrument thought to address gaps and unbalances after the pandemic, has been challenged by other rising shocks and stresses, which have contributed to its revision, such as geopolitical events, energetic issues, and social impacts of the envisaged transition with the risk to hamper the achievement of the ambitious goals of the EU Green Deal: decoupling economic growth from natural resources exploitation [2]. The EU Green Deal poses a challenge for the Union, where the economic development paradigm pursued so far has shown its main criticalities from the environmental perspective: overexploitation of natural resources, emissions and pollution, economic disparities, and social inequalities. The policy efforts deployed in the last decades have not addressed them successfully, given that territorial, social, and economic disparities persist [19] (if not widening in some cases). Then, the EU transition opens a window of opportunity to intervene for achieving sustainability and resilience from a multidimensional perspective. In this direction, the digital transition results are important if opportunely targeted to support the ecological one. Given their difference in nature, dynamics, and characteristics, it is extremely difficult to align them (ecological and digital) [25]. Instead, the aim is to exploit innovation and technological advancements to address environmental challenges, boost circular economic development, and support forms of social innovation and interactions between citizens, public, and private actors for a more inclusive society. Such an integrated approach finds in cities the ideal place to deploy possible innovative solutions [26, 27], and urban policies and planning can support such a transformative development.

Following this rationale, the book focuses on a better understanding of ESs in urban environments. It delves into the policy and practical aspects of ESs in cities and highlights the transformative potential of their integration into urban planning and design.

The first thematic area explores the role of ESs for and from the planning dimension, focusing on data-driven approaches. Cities today must innovate to thrive, and integrating ESs into urban planning can drive innovation. By leveraging interdisciplinary knowledge, cities can develop solutions that enhance sustainability and livability, addressing ecological and human needs. Cities and urban areas can be designed and managed to support ecological functions and provide multiple benefits to residents by improving green infrastructures, which are critical factors in innovative urban planning practices.

Indeed, green infrastructure and ESs are pivotal in fostering ecological transitions within cities and can contribute to creating resilient, adaptable, and ecologically sound cities. In this direction, the evolutionary nature of urban planning can contribute to address the challenges the challenges and opportunities of the transition.

The second thematic area covered by the book relates to data-driven approaches, frameworks, and methodologies for measuring ESs. It reflects the rising relevance of exploiting innovative technologies and approaches in our daily lives and elaborating innovative solutions to the challenges ahead. Such a also emerges in the urban planning field—and process—where the combination of interdisciplinary linkages between urban and territorial studies with computational science contributes to the development of innovative approaches and methods to better comprehend social, economic, and environmental phenomena and developing innovative approach in supporting decision-making and urban planning. Measuring ESs through data-driven approaches for urban planning is relevant for several reasons. Firstly, it provides vital information on the status and health of our natural capital, helping identify the level of natural ecosystem degradation and its ability to produce ESs. Obtaining precise and reliable information through data-driven approaches could enhance sustainability by helping to balance ecological, economic, and social goals. Moreover, by understanding the value of ESs, urban planners can effectively integrate natural capital into planning processes promoting sustainable development. This helps ensure that urban development does not come at the expense of environmental health and that cities can continue to provide essential services like clean air and water, climate regulation, and recreational spaces. Moreover, data-driven approaches in measuring ESs can support resilience and adaptation efforts. With climate change increasing the frequency and severity of extreme weather events, it is crucial to design urban areas that can withstand and recover from these impacts. By measuring ESs, planners can identify and enhance natural capital that provides critical buffering and adaptive functions, such as wetlands for flood control or urban green areas and forests for heat mitigation through green infrastructure design. Various methodologies for spatializing ESs and assessing their economic, social, and environmental benefits are exposed, providing a foundation for their inclusion in policy and planning. They emphasize the relevance of sustainability indicators as powerful tools for guiding urban development to implement green strategies, ensuring that urban growth aligns with ecological objectives. Moreover, such methods emphasize the importance of embracing an Adaptive Urban Planning perspective in facing rapid urban change and the potential of Big Data and Urban Informatics for Planning the Transition of Cities, delving into how these technologies can be harnessed to measure and manage ESs, facilitating the transition to more sustainable cities. At the same time, such thematic area inquiries also in the social dimension of the transition, addressing the socio-economic dimensions of urban ecological transitions in terms of green certifications and initiatives that can inadvertently contribute to spatial inequalities, highlighting the need for equitable planning approaches.

In conclusion, this book invites readers to explore the transformative potential of ecological and digital transitions in urban environments, offering practical insights and strategies for creating sustainable, resilient, and equitable cities.

The research projects at the core of contributions presented in this book represent an innovative effort to advance urban and territorial planning discourse and practice in the face of climate change. By integrating biodiversity, key enabling technologies, and inclusive economies, these projects offer a holistic and forward-looking approach essential for navigating contemporary urban challenges. The outcomes of these endeavors are poised to influence academic research, policy formulation, and urban planning practices, with the overarching goal of creating cities and territories that are sustainable, resilient, inclusive, and responsive to the evolving needs of diverse communities.

The contributions collected for this book have been presented at the International Symposium "Networks Markets and People - Communities, Institutions and Enterprises towards post-humanism epistemologies and AI challenges", scheduled from May 22–24, 2024, in Reggio Calabria, Italy, in the specific thematic sessions "Ecological And Digital Transition In Cities: Measuring Ecosystem Services For Urban Planning And Design", as part of the research activities conducted within three Next Generation EU-funded research projects (Pilot Project 4.6.1 – Goal 4.6 – Tech4You Innovation Ecosystems; ECO-SET—A Multidisciplinary approach to plan Ecosystem Services for cities in Transition; PLANET—PLANning Ecosystem services for cities in Transition). The conference saw the participation of high-quality international academics and experts from an international network of higher academic institutions by guesting significant contributions to stimulate a fruitful debate on global challenges among academics and policymakers. The themes discussed in these sessions followed the critical elements of the debate on a shift in policy design and implementation to drive transition-oriented structural changes in regions and cities. In this direction, this book offers the chance to navigate the complexity of transition and resilience by outlining possible policy agenda priorities, new approaches, cases, and experiences that enrich the flourishing academic and policymakers debate on the green and digital transition.

This volume as well as the design and organization of the thematic session is the result of the synergistic activities related to the following PNRR Research Projects funded by the European Union – NextGenerationEU: "Tech4You Innovation Ecosystem - Pilot Project 4.6.1 "Climate adaptation plans for the reduction of the ecological footprint and ecological debt, aimed at improving the conservation and transformation response in terms of resilience and quality of life in the Calabrian and Lucanian urban and territorial systems" (Goal 4.6 - Planning for Climate Change to boost cultural and natural heritage: demand-oriented ecosystem services based on enabling ICT and AI technologies), funded by the European Union – NextGenerationEU - PNRR - Mission 4 - Component 2 – Investment 1.5; the "ECO-SET - A Multidisciplinary approach to plan Ecosystem Services for cities in Transition" – project funded by the European Union – NextGenerationEU - PNRR - Mission 4 - Component 2 – Investment. 1.1 - (PRIN-PNRR); the "PLANET – PLANning Ecosystem services for cities in Transition - project funded by the European Union – NextGenerationEU - PNRR - Mission 4 - Component 2 - Investment. 1.2 (Young Researchers).

Francesca Moraci
Carmelina Bevilacqua
Pasquale Pizzimenti

References

1. European Union, "EU Strategic Agenda 2024-2029," https://www.consilium.europa.eu/en/european-council/strategic-agenda-2024-2029/.
2. EU Commission, "The European Green Deal. Communication From The Commission To The European Parliament, The European Council, The Council, The European Economic And Social Committee And The Committee Of The Regions.," 2019.
3. European Commission, "Nature Restoration Law Supporting the restoration of ecosystems for people, the climate and the planet," https://environment.ec.europa.eu/topics/nature-and-biodiversity/nature-restoration-law_en.
4. D. Haase, N. Frantzeskaki, and T. Elmqvist, "Ecosystem services in urban landscapes: Practical applications and governance implications," *Ambio*, vol. 43, no. 4, pp. 407–412, 2014, doi: 10.1007/s13280-014-0503-1.
5. A. Sirakaya, A. Cliquet, and J. Harris, "Ecosystem services in cities: Towards the international legal protection of ecosystem services in urban environments," *Ecosyst Serv*, vol. 29, pp. 205–212, Feb. 2018, doi: 10.1016/j.ecoser.2017.01.001.
6. M. Alberti *et al.*, "The Complexity of Urban Eco-evolutionary Dynamics," Sep. 18, 2020, *Oxford University Press*. doi: 10.1093/biosci/biaa079.
7. E. Gómez-Baggethun and D. N. Barton, "Classifying and valuing ecosystem services for urban planning," *Ecological Economics*, vol. 86, pp. 235–245, Feb. 2013, doi: 10.1016/j.ecolecon.2012.08.019.
8. P. Bolund and S. Hunhammar, "Ecosystem services in urban areas," 1999.
9. United Nations, "System of Environmental-Economic Accounting-Ecosystem Accounting White cover (pre-edited) version," Sep. 2021. [Online]. Available: https://seea.un.org/ecosystem-accounting.
10. Moraci F., Bevilacqua C., and Pizzimenti P., "Planning the transition of cities. Innovative research approaches and trajectories," *TeMA Journal of Land Use, Mobility and Environment*, vol. 1, pp. 109–127, 2024, doi: 10.6092/1970-9870/10430.
11. C. Cortinovis and D. Geneletti, "A framework to explore the effects of urban planning decisions on regulating ecosystem services in cities," *Ecosyst Serv*, vol. 38, Aug. 2019, doi: 10.1016/j.ecoser.2019.100946.
12. C., Cortinovis, D., Geneletti, B., Adem Esmail, and L. Zardo, *Planning for Ecosystem Services in Cities*. Springer, 2020. [Online]. Available: http://www.springer.com/series/8868
13. M. Scowen, I. N. Athanasiadis, J. M. Bullock, F. Eigenbrod, and S. Willcock, "The current and future uses of machine learning in ecosystem service research," *Science of the Total Environment*, vol. 799, Dec. 2021, doi: 10.1016/j.scitotenv.2021.149263.
14. A. Eraydin and T. Taşan-Kok, "Crisis in Planning Theory," in *Politics and Conflict in Governance and Planning*, Routledge, 2019, pp. 21–37. doi: 10.4324/9781351252881-2.
15. A. Eraydin, *Resilience Thinking in Urban Planning*, vol. 106. in GeoJournal Library, vol. 106. Dordrecht: Springer Netherlands, 2013. doi: 10.1007/978-94-007-5476-8.
16. E., Costa, C., Bevilacqua, C., Trillo, and G. Trupiano, "A new paradigm to conciliate neo-liberal economy and sustainable urban regeneration: Commercial Local Urban Districts," *PLANUM*, pp. 1–13, 2011.

17. L., Ricci, A., Iacomoni, and C. Mariano, "Nuova questione urbana e nuovo welfare. 1 Regole, strumenti, meccanismi e risorse per una politica Integrata di produzione di servizi," *ANANKE*, vol. 90, pp. 111–120, 2020.
18. Bevilacqua C. and Trillo C., "Urban Regeneration and Public-Private Partnership Issues. In "Economic Development Strategies. The Role of Public Private Partnership. First Scientific Report 2012 CLUDs Project.," 2012.
19. Bevilacqua C. and Pizzimenti P., "Resilient and Transition Strategies for the Post-pandemic City: A Multi-criteria Analytical Approach for the Case of the Metropolitan City of Reggio Calabria.," in *New Metropolitan Perspectives. NMP 2022. Lecture Notes in Networks and Systems, vol 639. Springer, Cham.* https://doi.org/10.1007/978-3-031-34211-0_8, C., Bevilacqua, PA., Balland, C., Kakderi, and V. Provenzano, Eds., 2023.
20. W. Rauws and G. De Roo, "Adaptive planning: Generating conditions for urban adaptability. Lessons from Dutch organic development strategies," *Environ Plann B Plann Des*, vol. 43, no. 6, pp. 1052–1074, Nov. 2016, doi: 10.1177/0265813516658886.
21. S. Kato and J. Ahern, "'Learning by doing': Adaptive planning as a strategy to address uncertainty in planning," *Journal of Environmental Planning and Management*, vol. 51, no. 4, pp. 543–559, Jul. 2008, doi: 10.1080/09640560802117028.
22. D. Crowley, T. Marat-Mendes, R. Falanga, T. Henfrey, and G. Penha-Lopes, "Towards a necessary regenerative urban planning. Insights from community-led initiatives for ecocity transformation," *Cidades*, pp. 83–104, 2021, doi: 10.15847/CCT.20505.
23. S. Schurig, "The concept of a 'regenerative city': How to turn cities into regenerative systems," 2021.
24. T. McPhearson, E. Andersson, T. Elmqvist, and N. Frantzeskaki, "Resilience of and through urban ecosystem services," *Ecosyst Serv*, vol. 12, pp. 152–156, Apr. 2015, doi: 10.1016/j.ecoser.2014.07.012.
25. S., Muench, E., Stoermer, K., Jensen, T., Asikainen, M., Salvi, and F. Scapolo, "Towards a green and digital future," Luxembourg, 2022. doi: 10.2760/977331.
26. E. Cosgrave, K. Arbuthnot, and T. Tryfonas, "Living labs, innovation districts and information marketplaces: A systems approach for smart cities," in *Procedia Computer Science*, Elsevier B.V., 2013, pp. 668–677. doi: 10.1016/j.procs.2013.01.070.
27. T. McPhearson, E. Andersson, T. Elmqvist, and N. Frantzeskaki, "Resilience of and through urban ecosystem services," *Ecosyst Serv*, vol. 12, pp. 152–156, Apr. 2015, doi: 10.1016/j.ecoser.2014.07.012.

Acknowledgements

Finanziato dall'Unione europea
NextGenerationEU

Italiadomani
PIANO NAZIONALE DI RIPRESA E RESILIENZA

Ministero dell'Università e della Ricerca

This edited volume is the result of the synergistic activities related to the following PNRR Research Projects funded by the European Union – NextGenerationEU: "Pilot Project 4.6.1 - Climate adaptation plans for the reduction of the ecological footprint and ecological debt, aimed at improving the conservation and transformation response in terms of resilience and quality of life in the Calabrian and Lucanian urban and territorial systems" project (Goal 4.6 - Planning for Climate Change to boost cultural and natural heritage: demand-oriented ecosystem services based on enabling ICT and AI technologies - Tech4You Innovation Ecosystem), Finanziato dall'Unione europea- Next Generation EU, Missione 4 Componente 2 - CUP C33C22000290006 (ECS_00000009); the "ECO-SET - A Multidisciplinary approach to plan Ecosystem Services for cities in Transition" project – Finanziato dall'Unione europea- Next Generation EU, Missione 4 Componente 2 - CUP B53D23029450001; the "PLANET – PLANning Ecosystem services for cities in Transition" project - Finanziato dall'Unione europea- Next Generation EU, Missione 4 Componente 2 - CUP C33C22000840006.

Contents

Data-Driven Perspectives for Unveiling Context Dynamics in Urban Planning and Design for the Transition

Adaptive Urban Planning for Sustainable Urban Transformations: A Data-driven Framework for Cities Ecosystem Services-Based Quantification

Francesca Moraci[1], Carmelina Bevilacqua[2], and Pasquale Pizzimenti[1](✉)

[1] DArTe Department, Università degli Studi Mediterranea, 89124 Reggio Calabria, Italy
pasquale.pizzimenti@unirc.it

[2] PDTA Department, LaSapienza Università Di Roma, 00196 Rome, Italy

Abstract. This paper provides a conceptual framework in supporting the definition of data-driven adaptive planning approaches for achieving sustainable urban transformations. The pressure on natural capital and the essential services it provides call for a transition of cities towards sustainability that can intertwine socio-ecological and socio-technological dynamics. Biodiversity, natural ecosystems, and ecosystem services come out as central themes for planning the transition of cities, and innovative approaches to urban planning and design are essential to tackle the issue of sustainability. Therefore, the study advocates for data-driven approaches that integrate purposefully selected indicators and metrics for land-use-derived categories that capture complex urban socio-ecological and socio-technological dynamics by integrating urban ecosystem services, urban planning and design, and urban climate indicators. This preliminary conceptual framework offers insights for urban planning that address current challenges and adapts to the dynamic nature of urban environments. Further investigations are needed to test the framework, data availability, and indicators in a case study area. With further methodological and empirical support, the proposed approach could serve as a valuable framework for adaptive urban planning strategies, guiding cities in their transition toward sustainability.

Keywords: adaptive planning · sustainable urban transformations · ecosystem services

1 Introduction

Urbanization processes are growing at a fast pace, increasing pressures on natural ecosystems[1–3]. Such processes are expected to continue [4]. By 2050, it is expected that 75% of the population will live in cities [1, 4–7]. The concentration of population in urban areas is posing challenges to human well-being [1], and at the same time, it generates an increasing demand for natural capital and ecosystem services in cities [5]. As the global landscape of urbanization continues to evolve, cities grapple with multifaceted challenges that demand innovative and adaptive solutions: biodiversity, climate

F. Moraci et al. (Eds.): NMP 2024, LNNS 1272, pp. 3–17, 2025.
https://doi.org/10.1007/978-3-031-82927-7_1

changes, energy issues, and poverty are calling for more sustainable cities [8]. Moreover, urbanization processes and their rapidly increasing pace are posing severe pressures on biodiversity and ecosystem services [2]. Recently, the focus of researchers and practitioners on creating sustainable and resilient cities highlights the importance of ecosystem services topic [4] and the relevance of their inclusion in practical implications, including spatial planning [9]. On the one hand, urbanization processes lead to densification, expansion/sprawling, and shrinking impacts on the environment [1]. On the other, given that urbanization is a complex form of social, economic, political, and technological factors, it offers the opportunity to understand better the interdependencies between impacts and responses to urbanization dynamics [1] that can increase the well-being of the population [2].

Despite the growing interest in the topic, both from the academic and the policy perspective, a preliminary literature investigation highlights a lack of attention from planning to ES [2] and a lack of consistency across policies on how to plan adaptation and monitor and assess its progress [10]. Efforts have been made regarding ESs assessment for their estimates and economic value, with specific attention to spatially explicit accounting frameworks [11]. However, despite the research on ES and urban planning addressing the specific performance of ES through planning and design, it is emerging the request for a common set of ecological indicators and metrics to assess their performance [4].

As cities expand, it becomes important to develop urban planning strategies that accommodate growth, prioritize environmental conservation, and enhance the quality of life for residents. An encouraging strategy involves integrating the quantification of ES, particularly in a spatially explicit manner, with valuation techniques [9].

Therefore, this paper seeks to find the answer to the following research questions: given the relevance of ESs identification and quantification in cities, also through spatially explicit methods, how adaptive urban planning can result crucial for achieving sustainable urban transformation for capturing the complexity of socio-ecological and socio-technological dynamics in cities through a data-driven planning perspective? And how is it possible to detect such a complexity from a data-driven perspective for achieving sustainable urban transformations through adaptive planning?

By exploring the concept of adaptive urban planning and its relevance in the context of sustainable urban transformations, the paper focuses on developing a conceptual framework to capture urban complexity through suitable indicators and metrics for promoting data-driven adaptive planning. The proposed conceptual framework, starting from the spatially explicit UN SEEA-EA approach for accounting ES in urban areas, is centered on three main conceptual data blocks: urban ecosystem services (UES), urban complexity, and urban climate zones. The first is rooted in the literature on UES measurement [4, 5], the second in the Ecosystemic Urbanism approach [12], the third in the literature on urban climate modeling [13]. Indicators and metrics have been framed and characterized on the land use categories following the common ES quantification frameworks [11] and eliminating possible data overlaps.

The paper is structured as follows. The next section explores the concept of ES and its relevance for the planning dimension from the adaptive urban planning perspective. The central section of the paper focuses on the conceptual indicators framework for planning sustainable transformations in cities, starting from their context conditions. The Discussion and Conclusions section outlines the conceptual framework proposal for data-driven adaptive planning.

2 Adaptive Urban Planning and Ecosystem Services

The literature review section underscored the pivotal role of spatial planning in advancing sustainable development. In response to the challenges posed by rapid industrialization and urbanization, the adaptive planning approach emerges as a central theme, addressing uncertainties associated with the transition of cities towards sustainability. Various terms, such as adaptive management, planning, and governance, converge in highlighting urban planning's distinctive characteristics in dealing with complexity and uncertainty. The review advocates for innovative approaches in urban planning and design to navigate sustainable land use amidst urbanization difficulties. Notably, the focus is on combining ecosystem services and spatial planning, as outlined in Ahern et al.'s transdisciplinary adaptive design planning model. This perspective underscores the importance of indicators and metrics in the quantitative approach to adaptive planning, emphasizing their role in evaluating and revising the goals to achieve.

Traditional planning, that tries to direct urban development towards a predetermined future spatial configuration is affected by uncertainties generating a mismatch between the desired, the intended and actual results [14]. In this direction, planning – in the era of complexity and uncertainties – should focus on the ability of cities to adapt to a "*dynamic*" environment rather than establishing a particular urban configuration [14]. In the context of the uncertainties to deal with in planning the transition of cities towards sustainability, the adaptive planning approach is coming to the fore [4, 15, 16]. The concept of Ecosystem Services holds significant promise for the adaptive planning discourse as it combines ecosystems functions and services with the flexibility to explore development paths in response of the complex challenges that cities are called to face [9]. Their relevance is also emphasized by their integration into urban planning at all levels: strategic-comprehensive, land-use zoning, and detailed development planning [17]. Central to the adaptive planning approach is the idea that it is not possible to define and predict an urban development trajectory's scenario rather it is possible to support a range of possible future configurations based on conditions for development, in which "possibility spaces" are generated and where urban structure can be shaped, and development processes unfold [14].

The literature sources examined present different terms related to the adaptive approach: adaptive management [16], adaptive planning [15], adaptive planning and design [4], adaptive governance [18]. All these approaches share some distinct characteristics of urban planning as well as the need to plan more sustainable cities starting from their ESs. Uncertainties are those "*unexpected natural, political, economic, events, coincidental confluences of gradual change processes feeding larger transformations and unforeseen societal responses to policy programmes*" [15]. Such uncertainties occur in multiple domains and at different scales and are connected with the development trajectories of cities, for example those related to the transformations induced by climate changes or technological innovations, represent a challenge for planners [15].

Adaptive approaches emerged at the end of 1970s with adaptive management concept with the aim to address the uncertainties inherent complex systems and manage it adaptively [16] starting from the recognition that ecosystems do not reliably revert to an equilibrium state after undergoing disturbance [19]. Adaptive management, rooted in natural resources management theory, entails an iterative decision-making process comprising a structured setup phase, specifying stakeholder involvement, objectives, actions, models, and monitoring plans, followed by an iterative phase that involves decision-making refinement, follow-up monitoring, assessment, and feedback, with widespread application across various policy contexts beyond public lands management [20]. For Rauws [15] an adaptive planning approach concentrates on enhancing urban areas' adaptability to anticipated and unexpected changes, striving to facilitate effective functioning under diverse circumstances. This approach, rooted also in complexity theory, necessitates a change in the emphasis of planning strategies, shifting from a focus on content (i.e., what) and process (i.e., with whom) to considerations of conditions for development [15] related to the capacity of city of transitioning to new configurations maintaining its system's robustness and livability over time (Rauws and DeRoo, 2016). Ahern et al. [4], have focused their inquiries onto the quantitative approach, in which indicators and metrics play an important role in the adaptive planning process. In their adaptive design framework, the identification of indicators and metrics for measuring goals, monitor and evaluate results, and applying findings, are crucial steps for defining goals related to UES for a specific plan and prioritize ESs goals and consider trade-offs and alternatives [4]. Although the interests and advancement of knowledge on the ESs topic, there is a difficulty to integrate them into regular urban planning [4], as well as practical tools to integrate social and biophysical information from a spatial planning perspective [9].

Therefore, the paper focused on those approaches that combine ESs and spatial planning in the definition of adaptive urban planning for sustainable urban transformations. An interesting attempt is provided by the Ecosystemic Urbanism developed by the Urban Ecology Agency of Barcelona (BCNEcologia-M) in 1995 [6] that introduces conceptual, methodological, organizational, and instrumental elements aimed at alleviating pressure on the support system and promoting the efficient enhancement of urban complexity [21]. This approach is deemed the most effective means to elevate the complexity of urban structures without unnecessary resource depletion, thereby restoring competitiveness to cities [21].

This preliminary literature analysis on the adaptive planning topic has emphasized the importance of integrated approach that goes beyond economic and social aspects, incorporating environmental dimensions and ecosystem-based solutions.

3 A Conceptual Framework for ES-Based Data-Driven Adaptive Planning

The starting point for the development of a conceptual framework aimed at measuring ecosystem services in cities for the development of data-driven planning for their transition is the spatially explicit United Nations System of Environmental and Economic Accounting – Ecosystem Services (UN SEEA - EA) [11]. In measuring urban ecosystem services, distinct challenges and limitations arise compared to other ecosystem types

[11]. One key consideration is the need for accurate change detection at the small spatial scales inherent in urban areas, where changes may be finer than the precision of land cover classifications used in ecosystem service models [11]. Spatial patterns in the supply of urban ecosystem services are influenced by biophysical variations, whereas spatial variations in demand, due to changes in population location and movement, may not be detectable at the same resolution[11]. The accounting of ES in urban ecosystem encompasses measures of their extent and data on their conditions (indicators and variables) and related ES [11]. Measuring ES in urban ecosystems offers the opportunity for a more detailed accounting according different urban areas sub-types that can provide useful information on urban green and blue spaces that can be provided for different boundaries and different spatial resolutions of statistical and reporting units. [11]. Indeed, such an account can be conducted in cities by "*administrative boundaries (i.e., local government boundary), functional boundaries (e.g., based on commuting flows as defined by census data), or morphological criteria, such as the extent of the built-up area plus a buffer zone* [11]. Urban areas range from high developed urban cores to less developed rural peripheral area often follow a gradient from less developed and even rural peripheral areas [11]. However, also in highly developed urban cores it is possible to find significant green areas [11]. The SEEA -EA suggests two possible approaches: the landscape approach and the individual asset approach [11] The first "*disaggregates the entire urban area and categorizes larger patches with common characteristics, classifying these areas according to different urban sub-types*" [11] and it is deemed relevant in supporting municipal planning and zoning [11]. The second detect information on individual asset type at high resolution through satellite imagines or other spatial datasets and allows for the identification of ecosystem assets in urban areas, for example allowing the identification of green and blue infrastructures able to provide ecosystem services [11].

The proposed conceptual framework (see Table 1) serves as a bridge between theoretical concepts and practical applications. By integrating indicators related to urban ESs, ecosystemic urbanism, and urban climatic zones, we provide a holistic perspective for evaluating the sustainability of urban transformations within the urban regeneration rationale. This multi-dimensional approach enables planners to consider not only the environmental aspects but also the social and economic dimensions of urban regeneration. The inclusion of ecosystemic urbanism principles further emphasizes the importance of a symbiotic relationship between human activities and the urban environment. The framework is articulated in three main indicators' groups.

The first refers to UES identification and measurement in cities. Quantifying UES in terms of extension and health status is important also for their economic accounting [11]. At this purpose we have outlined a set of indicators and metrics based on the work of Ahern et al. [4] and Gomez-Baggethun and Burton [5]. Therefore, the first indicators/metrics block refers to a set of indicators for measuring UES in urban settings that have been integrated in the work of Gómez-Baggethun & Barton [5], which provide a set of indicators framed according to the commonly accepted categorization of ecosystem services: provisioning, regulating, and recreative.

The second macro-category refers to the ecosystemic urbanism approach [22]. The approach is deemed relevant as it considers integrated dimensions in the analysis of the urban context for planning sustainable urban transformations. However, for the purpose of this study, the approach is considered as part of a wider rationale for the analysis of context conditions, that thanks to advantages deriving from emerging technologies, such as big data, machine learning and the urban informatics approaches, can generated

Table 1. Conceptual Framework indicators/metrics

Macro-category	Category	Indicator category	Indicator/metric	type of captial
Block 1- Urban Ecosystem Services Gómez-Baggethun and Barton (2013)Ahern et al. (2014)	provisioning	Food Supply	Production of food (tons yr − 1)	natural
		Food security	% urban green area dedicated to Agricultural activities, agricultural productivity	natural
		Habitat Provisioning	Index of Biotic Integrity (IBI), Fish Index of Biotic Integrity (FIBI), City Biodiversity Index (CBI)	natural
	Regulating	Water flow regulation and runoff mitigation	Soil infiltration capacity; % sealed relative to permeable surface (ha)	natural
		Water Quality	Total N, Total P, BOD, Turbidity, pH	natural
		Storm water infiltration	% impervious cover, soil permeability, slope of surface	natural
		Urban temperature regulation	Leaf Area Index; Temperature decrease by tree cover × m2 of plot trees cover (°C)	natural

(*continued*)

Table 1. (*continued*)

Macro-category	Category	Indicator category	Indicator/metric	type of captial
		Urban Climate	% tree canopy, maximum daily air temperature, diurnal heat flux	natural
		Noise reduction	Leaf area (m2) and distance to roads (m); noise reduction dB(A)/ vegetation unit (m)	natural
		Air purification	O3, SO2,NO2, CO, and PM10 μm removal (tons yr − 1) multiplied by tree cover (m2)	natural
		Air quality	Total particulates	
		Moderation of environmental extremes	Cover density of vegetation barriers separating built areas from the sea	natural
		Waste treatment	P, K, Mg and Ca in mgkg − 1 compared to given soil/water quality standards	natural
		Climate regulation	CO2 sequestration by trees (carbon multiplied by 3.67 to convert to CO2)	natural
		Carbon storage and sequestration	amount of carbon stored by urban trees	natural
		Pollination and seed dispersal	Species diversity and abundance of birds and bumble bees	natural
	Recreative	Recreation and cognitive development	Surface of green public spaces (ha)/inhabitant (or every 1000 inhabitants)	natural

(*continued*)

Table 1. (*continued*)

Macro-category	Category	Indicator category	Indicator/metric	type of captial
		Animal sightingt	Abundance of birds, butterflies and other animals valued for their aesthetic attributes	natural
		Public Recreation	Park visitation, activity mapping, "favorite places" identified by residents, and restorative experiences	social-relational
Block 2 - Ecosystemic Urbanism S. Rueda, B. Cormenzana, and M. Vidal, Ecological Urbanism Certifi-cation. Urbanism Certification with Sustainability Criteria,M.ofDe-velopment. Government of Spain, Ed., 2012	Compactness and Funcionality	Land Use	(net) Housing density	built
			Absolute compactness	built
			Corrected compactness	built
			Public open spaces per inhabitant	built
		Mobility and Services	Mode of transport of population	built
			Proximity to alternative transport networks	built
			Pedestrians area and spaces for coexistence	built
			Proximity to bicycle parking	built
			Off-street parking for vehicles	built
			Charging station for electric vehicle	built
			Labour market self-containment	built
		Public Spaces and Liveability	Air quality	built
			Acoustic comfort	built
			Thermal comfort	built
			Road accessibility	built

(*continued*)

Table 1. (*continued*)

Macro-category	Category	Indicator category	Indicator/metric	type of captial
			Liveability index in public space	built
	Complexity	Urban Organisation	Urban diversity index	built
			Density of legal persons	built
			Mix of uses	built
			Knowledge-dense activities	built
			Spatial and functional continuity of the street	built
		Green Spaces and Biodiversity	Biotic index of soil	natural
			Green space per inhabitant	natural
			Proximity to green spaces	natural
			Density of trees lining streets	natural
	Efficiency	Urban Metabolism	Residential energy consumption	built
			Services energy consumption	built
			Public lighting energy consumption	built
			Energy self-sufficiency. Renewables	built
			Greenhouse gas emissions	built
			Drinking water consumption	built
			Treated Wastewater	built
			Water sufficiency	built
			Waste generated per capita	built
			Gross separate collection	built

(*continued*)

Table 1. (*continued*)

Macro-category	Category	Indicator category	Indicator/metric	type of captial
			Closure of the organic material cycle	built
	Social Cohesion	Social Cohesion	Segregation index of the older population	social
			Segregation index of the foreign born population	social
			Segregation index of tertiary graduates	social
			Spatial/gender distribution of disposable	social
			household income	social
			Social inequalities rate	social
			Proximity to basic facilities	social
			Allocation of basic facilities	social
Block 3- Urban Heat Island	Local Climate Zones (LCZs) Stewart & Oke, 2012	Geometric properties (by LCZ)	Sky view factor	built
			Aspect ratio	built
		Surface properties (by LCZ)	Building surface fraction	built
			Impervious surface fraction	built
			Pervious surface fraction	built
			Height of roughness elements	built

(*continued*)

Table 1. *(continued)*

Macro-category	Category	Indicator category	Indicator/metric	type of captial
			Terrain roughness class	built
		Thermal properties by LCZ	Surface admittance	built
		Radiative properties by LCZ	Surface albedo	built
		Metabolic properties by LCZ	Anthropogenic heat output	built
Block 4 - Urban climate impacts indicators SIS (impact) indicators (EU Copernicus)	Urban SIS (impact) indicators (EU Copernicus)	Health	Air quality	built
		Energy	Energy Consumptions (Heating and cooling degree days) Solar energy (Solar insolation)	built
		Infrastructure	Flooding Soil Green infrastructure Transport infrastructure	built
		Non-specific sector	Temperature	built

data-driven decisions in planning the transition of cities through the adaptive planning perspective.

The last macro-category refers to the urban climatic conditions. Such a category is deemed important for the increasing relevance of climate changes in affecting planning decisions. The approach considers relevant climate variables useful to better prioritize planning choices in terms of adaptive interventions in the urban environment [23] and their characterization based on Local Climate Zones (LCZ) [13]. The resulting indicators/metrics have been selected based on their transformation into useful design parameters according to their relevance for the definition of adaptive planning strategies.

4 Discussion and Conclusion

The framework introduced in this study capitalizes on the principles of adaptive urban planning, emphasizing the need for flexibility and responsiveness in the face of dynamic urban challenges. Traditional urban planning approaches often struggle to accommodate the multifaceted nature of urban ecosystems and the intricacies of climate impacts. A data-driven framework harnesses the power of real-time data and advanced analytics to enable planners to make data-driven decisions that align with the evolving needs of urban environments. This adaptive approach not only enhances the resilience of cities but also fosters sustainable regeneration over time.

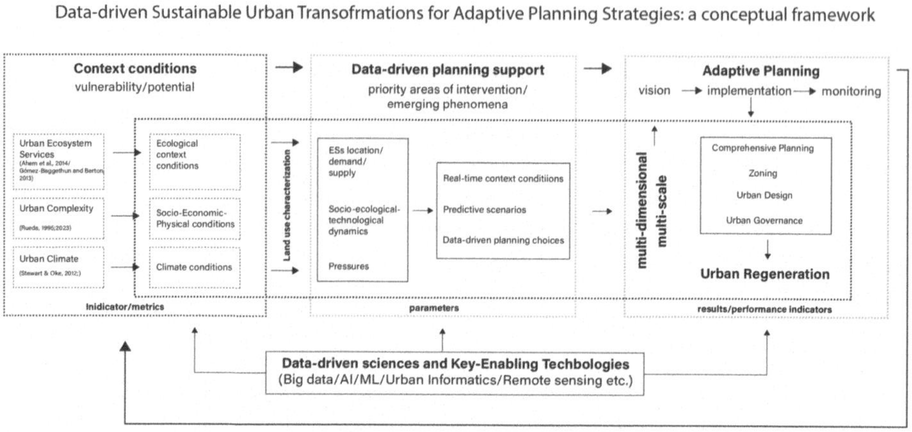

Fig. 1. Data-driven Sustainable Urban Transformations for Adaptive Planning Strategies: a conceptual framework. Authors' elaboration

The conceptual framework (see Fig. 1), starting from the analysis of complex context conditions (1st phase), can detect local vulnerabilities but also potentials given the aim to measure socio-ecological and socio-technological dynamics. Such activity, characterized by land-use categories, can provide useful data for quantifying ES demand/supply in cities and identify the area in which pressures are more evident (or areas more responsive to pressures) (2nd phase). Such data can measure real-time context conditions, generate predictive scenarios, and help formulate data-driven planning choices. In the 3rd phase, it would be possible to generate a vision for the transition of cities by characterizing urban planning tools in terms of comprehensive planning, zoning regulations, and urban governance processes. In the adaptive planning perspective, the monitoring phase is a crucial one. The availability of data and the clear definition of goals and indicators [4], merged with the emerging technologies in the big data field, can provide real-time monitoring conditions to assess the impacts of planning choices and, in case of emerging negative impacts to reformulate them for a better outcome.

The proposed framework offers a potential systematic and quantifiable means of assessing the context conditions for developing a range of possible scenarios useful for urban planning decisions on ESs. The framework provides a set of indicators for cities to embrace adaptive strategies that enhance the quality of life for residents while minimizing negative environmental impacts.

While this study lays the groundwork for adaptive urban planning approaches it presents some limitations. One of the intrinsic limitations is related to the availability of data. Therefore, further work should investigate available data on selected case study areas to refine (expand or reduce) the indicators/metrics list and possibly verify, in case of lack of data, which can be retrieved by exploiting big data methods. Also, it has to be noticed that many of indicators/metrics in the different macro-categories depicted in the

Table can overlap. While it can appear redundant, and this could be confirmed by further investigations, these indicators/metrics are used for different purposes, and therefore it could represent a potential in terms of using the same indicator/metrics for detecting different context conditions by transforming them in different design parameters. At the same time, future studies focusing on refining and expanding the indicator set can consider and verify such overlaps by selecting, in case, the proper indicators/metrics that respond to the rationale of the framework. Finally, integrating emerging technologies, such as artificial intelligence and machine learning, could enhance the framework's potential. Exploiting such technologies could generate interesting implications for the entire planning process. It could accelerate the gathering of data needed to build the knowledge framework about the context conditions, unveil non-linear and emerging phenomena related to the complexity of socio-ecological-technological dynamics, support the decision-making system with precise spatial-temporal information, allow real-time monitoring of transformations increasing the flexibility of planning implementation tools.

In conclusion, the data-driven conceptual framework presented in this paper represents a preliminary step toward a more sustainable and resilient urban future. By embracing adaptive planning principles and incorporating a comprehensive set of indicators, cities can navigate the complexities of urban transition with a heightened awareness of their impact on ecosystem services.

Acknowledgements. This research work is the result of the synergistic activities related to the following PNRR Research Projects funded by the European Union – NextGenerationEU: "Pilot Project 4.6.1 - Climate adaptation plans for the reduction of the ecological footprint and ecological debt, aimed at improving the conservation and transformation response in terms of resilience and quality of life in the Calabrian and Lucanian urban and territorial systems" (Goal 4.6 - Planning for Climate Change to boost cultural and natural heritage: demand-oriented ecosystem services based on enabling ICT and AI technologies - Tech4You Innovation Ecosystem), Finanziato dall'Unione europea- Next Generation EU, Missione 4 Componente 2 - CUP C33C22000290006 (ECS_00000009); the "ECO- SET - A Multidisciplinary approach to plan Ecosystem Services for cities in Transition" project – Finanziato dall'Unione europea- Next Generation EU, Missione 4 Componente 2 - CUP B53D23029450001; the "PLANET – PLANning Ecosystem services for cities in Transition" project - Finanziato dall'Unione europea- Next Generation EU, Missione 4 Componente 2 - CUP - C33C22000840006.

References

1. Haase, D., Frantzeskaki, N., Elmqvist, T.: Ecosystem services in urban landscapes: practical applications and governance implications. Ambio **43**(4), 407–412 (2014). https://doi.org/10.1007/s13280-014-0503-1
2. Sirakaya, A., Cliquet, A., Harris, J.: Ecosystem services in cities: towards the international legal protection of ecosystem services in urban environments. Ecosyst. Serv. **29**, 205–212 (2018). https://doi.org/10.1016/j.ecoser.2017.01.001
3. Alberti, M., et al.: The Complexity of Urban Eco-evolutionary Dynamics. Oxford University Press (2020). https://doi.org/10.1093/biosci/biaa079

4. Ahern, J., Cilliers, S., Niemelä, J.: The concept of ecosystem services in adaptive urban planning and design: a framework for supporting innovation. Landsc. Urban Plan. **125**, 254–259 (2014). https://doi.org/10.1016/j.landurbplan.2014.01.020
5. Gómez-Baggethun, E., Barton, D.N.: Classifying and valuing ecosystem services for urban planning. Ecol. Econ. **86**, 235–245 (2013). https://doi.org/10.1016/j.ecolecon.2012.08.019
6. Cormenzana, B., Fabregas, F., Marinescu, M.-C., Marrero, M., Rueda, S., Uceda-Sosa, R.: An Ontology for Ecological Urbanism: SUM+Ecology (2014). www.aaai.org
7. Wolfram, M., Borgström, S., Farrelly, M.: Urban transformative capacity: from concept to practice. Ambio **48**(5), 437–448 (2019). https://doi.org/10.1007/s13280-019-01169-y
8. Salat, S., Bourdic, L.: Urban complexity, efficiency and resilience. In: Energy Efficiency - A Bridge to Low Carbon Economy, InTech (2012). https://doi.org/10.5772/38599
9. Lerouge, F., Gulinck, H., Vranken, L.: Valuing ecosystem services to explore scenarios for adaptive spatial planning. Ecol. Indic. **81**, 30–40 (2017). https://doi.org/10.1016/j.ecolind.2017.05.018
10. Olazabal, M., Ruiz De Gopegui, M.: Adaptation planning in large cities is unlikely to be effective. Landsc Urban Plan **206** (2021). https://doi.org/10.1016/j.landurbplan.2020.103974
11. United Nations. System of Environmental-Economic Accounting-Ecosystem Accounting White cover (pre-edited) version (2021). https://seea.un.org/ecosystem-accounting
12. Rueda, S.: Ecological Urbanism. Barcelona (2023). https://doi.org/10.5040/9781350394612.ch-007
13. Stewart, I.D., Oke, T.R.: Local climate zones for urban temperature studies. Bull. Am. Meteorol. Soc. **93**(12), 1879–1900 (2012). https://doi.org/10.1175/BAMS-D-11-00019.1
14. Rauws, W., De Roo, G.: Adaptive planning: generating conditions for urban adaptability. lessons from Dutch organic development strategies. Environ. Plann. B. Plann. Des. **43**(6), 1052–1074 (2016). https://doi.org/10.1177/0265813516658886
15. Rauws, W.: Embracing uncertainty without abandoning planning: exploring an adaptive planning approach for guiding urban transformations. DISP **53**(1), 32–45 (2017). https://doi.org/10.1080/02513625.2017.1316539
16. Kato, S., Ahern, J.: 'Learning by doing': adaptive planning as a strategy to address uncertainty in planning. J. Environ. Planning Manage. **51**(4), 543–559 (2008). https://doi.org/10.1080/09640560802117028
17. Cortinovis, C., Geneletti, D., Hedlund, K.: Synthesizing multiple ecosystem service assessments for urban planning: a review of approaches, and recommendations. Landsc Urban Plan **213**, 104129 (2020). https://doi.org/10.1016/j.landurbplan.2021.104129
18. Green, O.O., et al.: Adaptive governance to promote ecosystem services in urban green spaces. Urban Ecosyst **19**(1), 77–93 (2016). https://doi.org/10.1007/s11252-015-0476-2
19. Birgé, H.E., Allen, C.R., Garmestani, A.S., Pope, K.L.: Adaptive Management for Ecosystem Services (2016)
20. Ruhl, J.B.: Adaptive management of ecosystem services across different land use regimes. J. Environ. Manage. **183**, 418–423 (2016). https://doi.org/10.1016/j.jenvman.2016.07.066
21. Cormenzana, B., Marinescu, M.C., Marrero, M., Mendoza, S., Rueda, S., Uceda-Sosa, R.: Models for sustainability. In: 2015 IEEE 1st International Smart Cities Conference, ISC2 2015, Institute of Electrical and Electronics Engineers Inc. (2015). https://doi.org/10.1109/ISC2.2015.7366221
22. Rueda, S.: Certificación del urbanismo ecosistémico (2012). http://www.bcnecologia.net/sites/default/files/publicaciones/docs/certif_urb_ecosistemico_web.pdf
23. Sharma, R., Hooyberghs, H., Lauwaet, D., De Ridder, K.: Urban heat island and future climate change—implications for delhi's heat. J. Urban Health **96**(2), 235–251 (2019). https://doi.org/10.1007/s11524-018-0322-y

Data-Driven Planning Approaches for Measuring Urban Ecosystem Services: The Potential of Big Data and Urban Informatics for Planning the Transition of Cities

Francesca Moraci[1], Carmelina Bevilacqua[2], Pasquale Pizzimenti[1](✉), Vincenzo Barrile[3], and Emanuela Genovese[3]

[1] DArTe Department, Università degli Studi Mediterranea, 89124 Reggio Calabria, Italy
pasquale.pizzimenti@unirc.it
[2] PDTA Department, La Sapienza Università di Roma, 00196 Rome, Italy
[3] DICEAM Department, Università degli Studi Mediterranea, 89124 Reggio Calabria, Italy

Abstract. This paper explores data-driven planning approaches for measuring urban ecosystem services in cities by focusing on the potential of big data. In the face of the ecological and digital transition, technologies' advancements, big data, and innovative analytical approaches offer novel perspectives on capturing the dynamic interactions within urban ecosystems and contribute to the definition of data-driven planning approaches for the transition of cities. The paper outlines possible approaches to integrating diverse open data sources to gain comprehensive insights into urban ecosystem services measurement in cities by exploiting satellite images and remote sensing technologies in a case study area. For this purpose, two commonly used vegetation indexes have been used, namely the Normalized Difference Vegetation Index and the Leaf Area Index, combined with Land Use data following the urban informatics approach. Results show the potential of data-driven approaches for planning the transition of cities towards sustainability and resilience and the implications of such data and technologies for urban planning and governance.

Keywords: data-driven planning · transition · big data · urban informatics

1 Introduction

This paper explores data-driven planning approaches for measuring urban ecosystem services, focusing on the potential of big data and urban informatics. The concentration of population in urban areas is constantly rising, together with the demand for natural resources [1, 2]. The pressure on ecosystem services (ES) is increasing in cities [3] which are expected to ensure favorable living conditions for the global population. Such a concentration of people confers a high level of dynamism and functionality to urban spaces [4], outlining the complexity of cities and their ability to be catalysts for transformative development. Cities and urban areas play a crucial role in global sustainability, serving

F. Moraci et al. (Eds.): NMP 2024, LNNS 1272, pp. 18–34, 2025.
https://doi.org/10.1007/978-3-031-82927-7_2

as focal points for sustainability advancements and catalysts for global transformation, particularly in areas such as energy efficiency, climate change adaptation, and social innovation [5]. At the same time, cities depend on ecosystems beyond their limits and benefit from internal urban ecosystems [1]. This aspect introduces the concept of Urban Ecosystem Services (UES), which is becoming influential in planning, designing, and managing cities toward sustainability [6, 7]. UES enhances the quality of urban life, as locally generated services like air quality and noise levels contribute significantly to the improved quality of life compared to services dependent on distant ecosystems [1, 8]. Therefore, detecting and accounting for ecosystem assets and services in urban areas becomes a crucial aspect to investigate [8].

Despite the advancement of knowledge on the contribution of UES to the health and well-being of urban residents, they still need to be sufficiently integrated into urban governance and planning [5]. In addition, data crucial for ecosystem services (ES) research, such as remotely sensed data, often exhibits high dimensionality, lacks a clear structure, and is accumulating at a pace that surpasses our capacity to interpret it through conventional methods [9]. Moreover, some scholars [10] have highlighted how new methodologies and technologies for assessing ES in scale and resolution are needed to match urban planning actions better, such as urban remote sensing with spatial big data [4]. Therefore, this contribution explores the connection between UES and data-driven Planning, highlighting the innovative opportunities presented by utilizing large-scale data and urban informatics approaches, and seeks to answer the following research question: 1. How can the integration of big data, new technologies and innovative approaches, such as urban informatics, enhance the assessment and understanding of urban ecosystem services, particularly in relation to their role in promoting sustainable urban development and human well-being?

The methodology describes a data-driven process characterized by a Landscape Approach [8] in measuring UES by detecting and measuring two common vegetation indexes in a test-case study area as an example of how big data and urban informatics can help measure UES in cities: the Normalized Difference Vegetation Index (NDVI) and the Leaf Area Index (LAI).

The paper is structured as follows. The next section presents a preliminary literature review of the main concepts beyond this study: the relevance of big data and urban informatics approaches for data-driven planning, and the role of emerging technologies and big data for measuring UES through remote sensing technologies. The central section of the paper describes the overall methodology outlining the potential of urban informatics in measuring UES for data-driven planning through the detection of NDVI and LAI and their mapping in relationship to land use categories. The discussion and conclusion section outlines this contribution's limitations, the main implications of such approaches for the urban planning and governance dimension, and the potential future research trajectories to explore.

2 Background

2.1 Big Data and Urban Informatics for Data-Driven Planning Approaches

The rise of big data presents opportunities to evaluate people's interactions with urban nature, helping to better comprehend utilization, significance, and social fairness in accessing specific spaces in the city that contribute to human well-being [11]. It is coming to the fore the transition into the era of big data, where information pertaining to understanding and managing cities is rapidly expanding into an overwhelming flow of "timely, varied, resolute and relational data" [12]. The term "big data" is not new as it was used in the early 1990s [4], and despite the growing interest on the concept in the last decade, it still lacks a clear definition [13], with different interpretations in many disciplines and user communities [14]. Kitchin [15] has outlined the main characteristics of Big Data from an analysis of the literature: "huge in volume; high in velocity; diverse in variety; exhaustive in scope; fine-grained in *resolution* and uniquely *indexical* in identification; *relational* in nature; *flexible*, holding the traits of *extensionality* and *scaleability*. However, multiple forms of Big Data exist, and not all data types could present all the commonly understood Big Data characteristics or require complex statistical techniques and computational capability for their analysis [16]. At the same time, the amount of data in urban areas generated from citizens, their interaction with technological devices, and traditional data sources such as censuses, provide the opportunity for a better management of the urban environment and transform the city in a data-rich environment [2], and complementing them with traditional data (survey, census, etc.) increases the analytical capacity in supporting planning [17]. In this context, the urban informatics approach is coming to the fore, defined as "the study of urban phenomena through a data science framework of urban sensing, data mining and integration, modeling and analysis, and visualization to generate new insights that simultaneously advance methods in computational science and address domain-specific urban challenges" [18]. Thakuriah et al. [14] conceptualize Urban Informatics as the exploration and comprehension of urban patterns and processes. O'Brien [13] has defined urban informatics as a relatively young discipline that uses "*digital technology and data to better understand and serve communities*" in the context of the proliferation of generated data and digital tools for their collection, organization, and analysis. Batty [19], in defining urban informatics as provided three main thematic strands related to the concept: exploiting technological devices (computers) and advancements (computation) across spatial and temporal domains, and from the interactions between humans and technology; modelling and simulation for making sense of big data and increase cities efficiency; the contribution of such activities for modeling and simulating in the development of a "*science of cities*" centered on the combination of big data and computation for the development of theories and methods useful for a better understanding of cities [19].

The availability of Big Data and the adoption of novel research practices and approaches is promoting the emergence of new research paradigm characterized by data-intensive exploration in which two main drivers can be distinguished for a shift towards data-driven planning: a first one in which data exploration is independent from theory, and a second one characterize by data-driven science which entails a modification of the existing scientific method [15].

Given this context, the intersection of urban informatics and big data can result increasingly pivotal in advancing data-driven approaches for measuring UES in cities. As cities become more instrumented and networked, vast quantities of high-resolution temporal and spatial data are collected rapidly through sensors on satellites, planes, and drones or simply by the interactions between humans and technological devices. Therefore, the next section investigates the linkages between big data and technologies for the UES measurement in cities.

2.2 The Role of Emerging Technologies and Big Data for Measuring Urban Ecosystem Services: Remote Sensing Approach

The literature on ecosystem services (ESs) is growing with an increasing number of studies focusing on their modeling and mapping from the local to the global scales [20] requiring the combination of diverse data on both ecosystems and the economy [21]. Thanks to the use of advanced technologies and big data, data-driven urban planning approaches centered on UES could contribute to optimize environmental, social, and economic benefits. Data-driven approaches exploit data for the definition of a knowledge base useful to detect problems, identify priorities, define and implement a policy, monitor its efficacy, and the definition of an eventual adaptive response [22]. Data-driven planning methodologies offer the potential for a better comprehending of the dynamics of urban ecosystem services, specifically the collection of new data and the exploitation of advanced technologies, such as remote sensing could contribute to gather comprehensive data and complement the existing ones [23].

Specifically, accurate geospatial data are essential to detect and depict their distribution and condition for modeling the flow of ecosystem services. [21] For example, the application of remote sensing technologies in urban areas is commonly applied for urban ecosystem services analysis, and the potential of spatial big data in urban studies is coming to the fore [4]. The utilization of emerging technologies has the potential to enhance ES assessment tools, in terms of [24]: 1) Higher resolution local input data; 2) Data science tools for locally fit transfer functions; 3) Insights in trade-offs and synergies among ecosystem services. The first refers to the exploitation of high-resolution spatial-temporal data, which allows for the estimation of ecosystem services at a local scale, aligning with the decision-making context of many local and regional authorities [24]. The second refers to the exploitation of data science for a better refinement of ES transfer functions [24]. The third, refers to the digitalisation of ES calculations, which facilitates overlaying the spatial occurrences of ES [24].

The remote sensing field is experiencing rapid development, characterized by the swift collection of substantial volumes of high-resolution temporal and spatial data gathered more rapidly than ever before through sensors mounted on various platforms, including satellites, planes, and recently, drones [14].

High resolution spatial data are important for ESs accounting in urban areas [8]. The spatially explicit United Nations System of Environmental and Economic Accounting – Ecosystem Services (UN SEEA - EA) [8] emphasized the need for accurate measurement at the small spatial scales inherent in urban areas. This relevance is due both for the influence of biophysical variations on the spatial patterns in the supply of ES in urban areas and also to the spatial variations in demand for ES due to population dynamics, which can be difficult to detect at the same resolution [8] The UN SEEA – EA [8] points out the different accounting processes for ES in urban areas, that can follow administrative boundaries, functional boundaries or morphological criteria. The UN SEEA – ES [8], following the morphological criteria suggests two possible approaches that can fit with rationale of the paper: the landscape approach and the individual asset approach [8] The first "*disaggregates the entire urban area and categorizes larger patches with common characteristics, classifying these areas according to different urban sub-types*" [8]. The second detects information on individual asset type at high resolution through remote sensing technologies or other spatial datasets, and allowing for the detection of ecosystem assets in urban areas [8].

Many authors and ES accounting frameworks focus on measuring biophysical vegetation parameters better to understand natural ecosystem extent variations and health status [8, 25, 26]. Two common vegetation indexes used for this purpose are the Normalized Difference Vegetation Index (NDVI) and the Leaf Area Index (LAI). The NDVI (Normalized Difference Vegetation Index) obtained from remote sensing is a widely used index to evaluate the health and density of vegetation on a large scale [27] The NDVI is based, as it is known, on the normalized difference between the reflectance in the near-infrared (NIR) band and the red (RED) band of the electromagnetic spectrum. In the literature, many studies have investigated the relationship between the Normalized Difference Vegetation Index (NDVI) and the Leaf Area Index (LAI) [28]. NDVI and LAI are widely used indicators for determining biomass [29]. The LAI is an important biophysical variable - adopted as required input for various ecological models - that outlines photosynthesis, transpiration, and energy balance, and it is defined "as the projected area of leaves per unit of ground area" [28]. Measurement of LAI can occur directly or indirectly, and, since in situ measurement of LAI can be expensive and often impractical for remote locations, there is growing interest in using remote sensing data to estimate this index [28]. Previous research has demonstrated a strong correlation between the red-to-near infrared transmission ratio and the LAI [28]. Leaves have different optical characteristics in the visible and near-infrared wavelengths. High absorbs leaf pigments due to chlorophyll and carotenoids lead to low reflectance in the visible range [28]. In contrast, plant cell walls cause near-infrared energy dispersion, resulting in relatively high near-infrared transmission and reflectance [28]. The use of spectral measurements is therefore related to LAI [28]. This has led to the development of empirical algorithms that link LAI to surface reflectance and reflectance-derived vegetation spectral indices [28].

The utilization of satellite data, as evidenced by the extraction of NDVI (Normalized Difference Vegetation Index) and LAI (Leaf Area Index) from such data sources, facilitates the differentiation of vegetation, enables a quantitative evaluation of its growth stage, and allows for estimates of carbon absorption and evapotranspiration allowing for useful insights in the quantification of ES, also in cities [30].

Given the relevance of these metrics for the potential calculation of UES, their possible calculation using remote-sensing data and technology, and the possibility of combining them with land-use data and other statistical data, they have been considered useful for providing an example of the potential of big data and urban informatics as a data-driven approach for measuring UES.

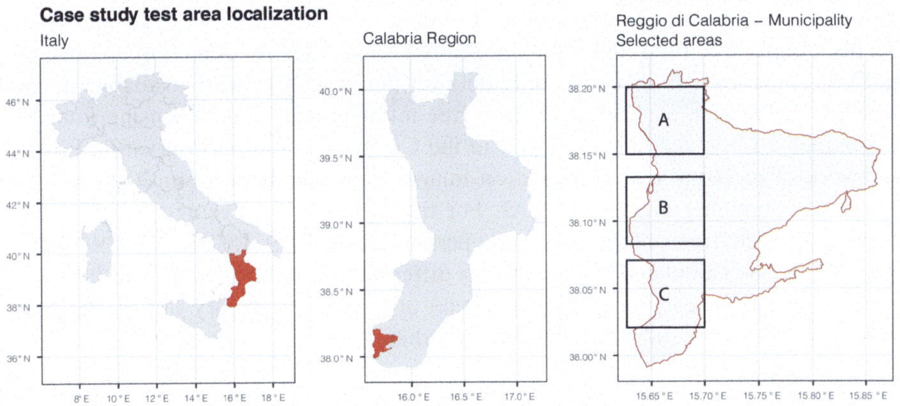

Fig. 1. Case study test area: the Municipality of Reggio Calabria, Italy and Areas selected for the combination of NDVI and LAI data with Land Use data Source: Author's elaboration

3 Methodology

The guiding principle of the methodology was to provide an example of technologies and data exploitation for a data-driven approach to measuring UES. Such an example is rooted in the application of remote sensing technologies for estimating NDVI and LAI from remotely sensed images and then aggregating such data for driving urban planning. The study has focused on the test case of the Reggio Calabria municipality, located in southern Italy – Calabria Region - and consisted of the following steps. First, data availability regarding satellite images on the case-test area has been assessed from the EU Copernicus Systems supplemented with Land Use data. Second, data extracted were used to calculate the NDVI and LAI averages in the two years selected (2019 and 2023), obtaining their variation. Third, given the amount of data to process, three small areas within the case-test area have been selected based on the high percentage of urbanized areas. Fourth, NDVI and LAI have been associated with each Land Use parcel to detect their change in urban areas.

3.1 Case Study Area and Data

The city of Reggio Calabria is located in Southern Italy in the Calabria Region (See Fig. 1), and it is the municipality with the higher population and surface in the regional context. The city of Reggio Calabria has a population of roughly 170.000 inhabitants. The Municipality, with an extension of 242.54 sq km, ranks among the largest municipalities in Italy in terms of size. Although the population has not shown significant variations in recent decades, it has confirmed the trend, demonstrated for several years, of preferring hilly and coastal areas in the southern periphery as settlement areas, contrary to the predictions of the City plan that indicated the coastal and hilly axis of the northern periphery as the direction of expansion[1]. For this study, two different open data sources have been exploited. First, the EU Copernicus Urban Atlas for detecting and mapping Land Use categories. In this case, only data for the year 2018 were available. Second, following Kollányi et al. [29], data from free multispectral remote sensing images of the Sentinel-2 L2A satellite developed in the Copernicus Earth Observation Program framework. The 10 m spatial resolution images were available from 2016 and were suitable for obtaining NDVI and LAI. For the analysis, the average for the images from two different times of the vegetation period (20/04/2019, 15/10/2019, 29/04/2023, 11/10/2023) was calculated for about two different years. The NDVI and LAI value averages have been calculated for each type of land use category during the period considered (2019 and 2023) for the selected urban areas.

4 Results

The first step in the methodology involved assessing data availability for the Reggio Calabria municipality. In this phase, land use data from the EU Copernicus Urban Atlas was detected and mapped for the Municipality of Reggio Calabria (Fig. 2a–2b). By exploiting the EU Urban Atlas 2018 land use data, it has been possible to provide insights into the municipality land use (see Fig. 2). Figure 2 shows how forests and pastures cover a high percentage of the municipality surface, with 24.43% and 24.81%, respectively. Arable land and Herbaceous vegetation associations cover 19.97% and 11.12%, respectively. All these categories represent 80.33% of the surface covered, indicating that only 19.27% is covered by urban fabric, infrastructures, and "other" categories. Considering the urban fabric categories, their surface covers 4.84%, with other "urban" categories (such as roads, infrastructures, industrial, commercial areas, etc.) covering 1.8%. At first glance, the analysis of land use coverage at the municipal level unveils an overall potential of ES for the entire municipality in terms of quantification of provisioning, regulating, and supporting services. Forests, vegetation, and arable land and pastures contribute to the local production of ES. However, such data have offered only a preliminary insight into the ecosystem's extent (Fig. 3).

Following the gathering of satellite imagery and Land Use data, the next phase involved extracting relevant information to calculate NDVI and LAI averages. Two specific years, 2019 and 2023, were selected for analysis. This step aimed to capture the temporal variation in NDVI and LAI, providing insights into the changes in biomass over the specified period (See Fig. 4a, 4b, 5a,5b). This methodological step highlights the potential of remote sensing data and technology in providing useful high spatial and temporal resolution data.

Due to the substantial volume of data, a strategic approach was adopted by selecting three smaller areas within the Reggio Calabria municipality. These areas were chosen based on a high percentage of urbanized land. These areas have been selected according to the presence of more dense urban fabric. The first (see Area A in Fig. 1c) is the city's urban core, characterized by a highly dense urban fabric with the historical center and the consolidated city. The second (see area B in Fig. 1c) is located in the southern

[1] Technical report for the Municipality Comprehensive Plan (PSC, 2019).

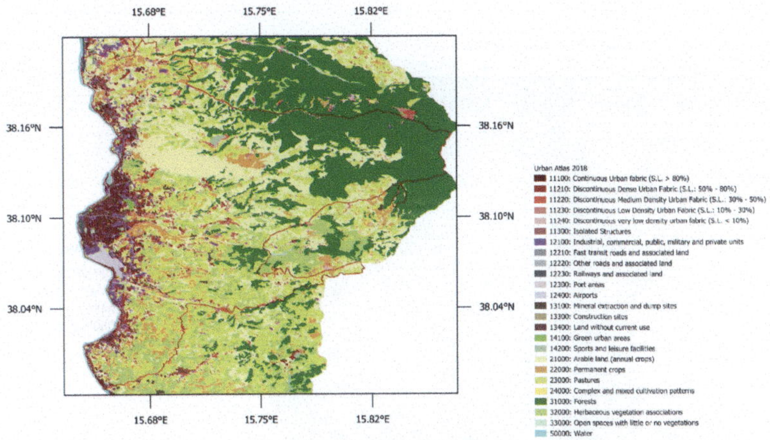

Fig. 2. Case study test area Land Use map: the Municipality of Reggio Calabria, Italy. Source: Author's elaboration based on EU Copernicus Urban Atlas 2018

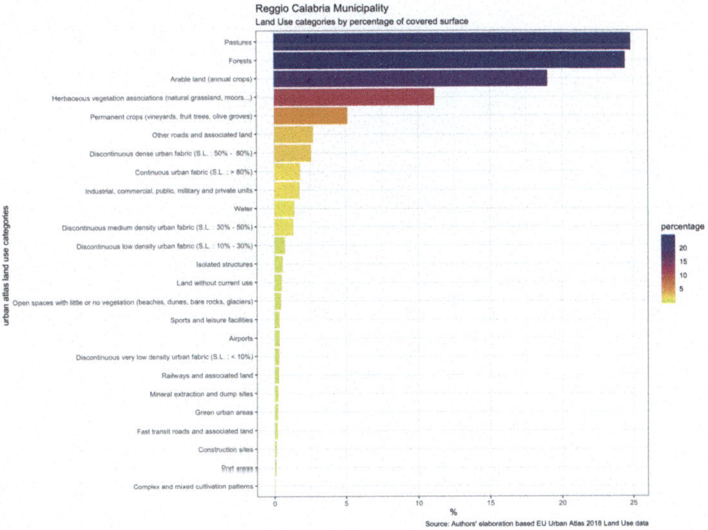

Fig. 3. Case study test area Land Use data: the Municipality of Reggio Calabria, Italy. Source: Author's elaboration based on EU Copernicus Urban Atlas 2018

part of the municipality, a consolidated suburb with a small urban center surrounded by industrial and agricultural areas. The third (see area C in Fig. 1c) is an area in the northern part of the municipality, a consolidated suburb characterized by two small centers and a more fragmented urban morphology. This segmentation facilitated a more detailed and focused analysis, enhancing the precision of the results. The NDVI and LAI values obtained were then associated with each Land Use parcel within the selected urban areas. This association aimed to identify and quantify changes in NDVI and LAI, specifically

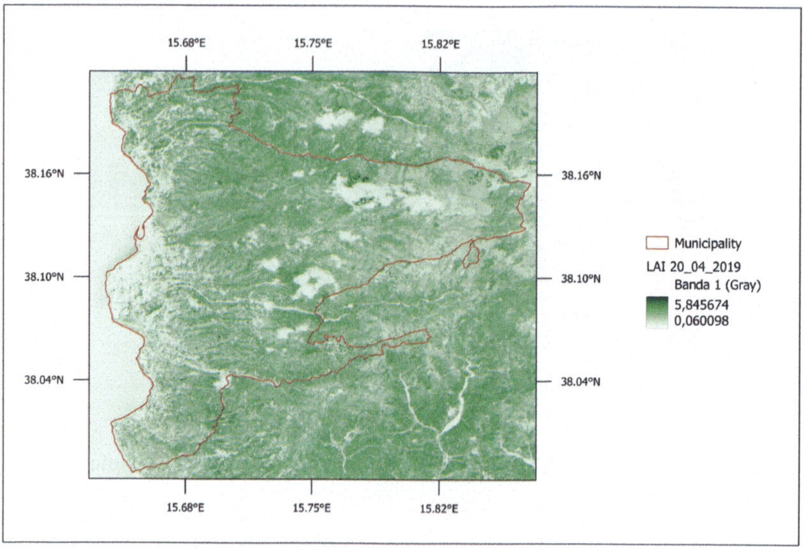

(a)

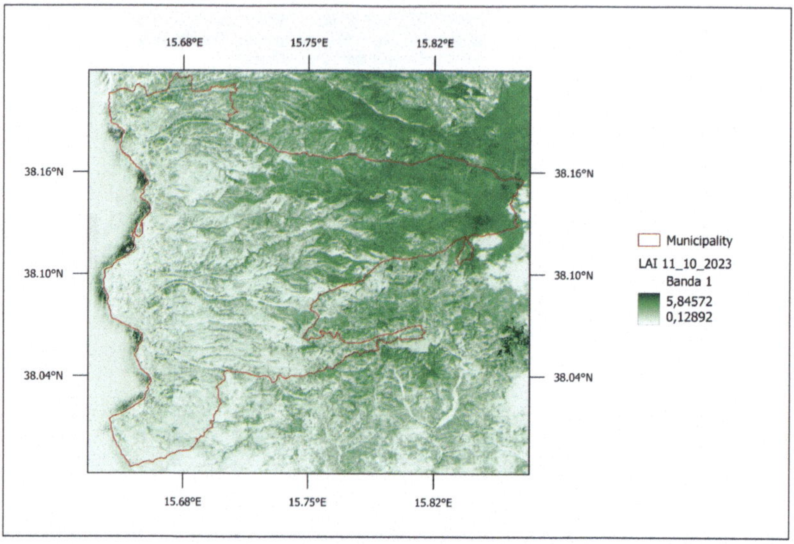

(b)

Fig. 4. Municipality of Reggio Calabria average Leaf Area Index 2019 (a) and 2023 (b). Source: Author's elaboration

within urban regions. By linking these indicators to land use categories, the study sought to highlight variations in environmental sustainability within different urban zones.

The final step involved the analysis of NDVI and LAI changes within the identified urban areas (see Figs. 6a, 6b, 7a, 7b, 8a, 8b). In these areas, Land Use data have

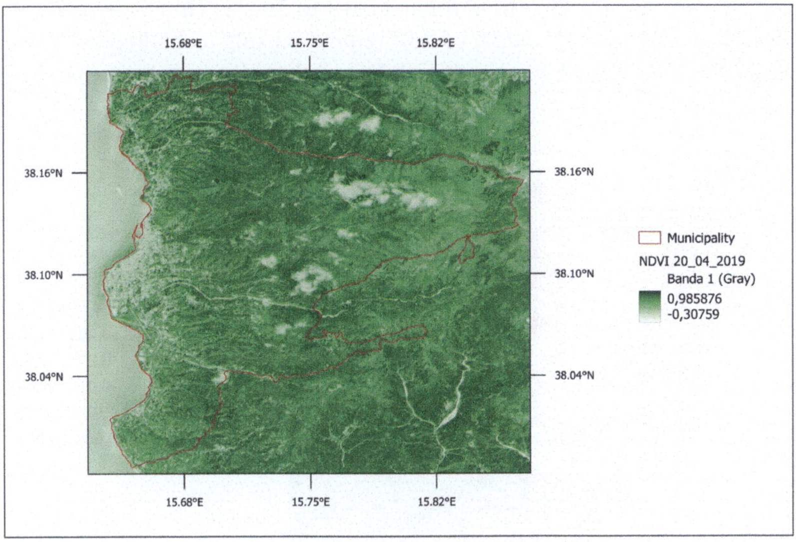

(a)

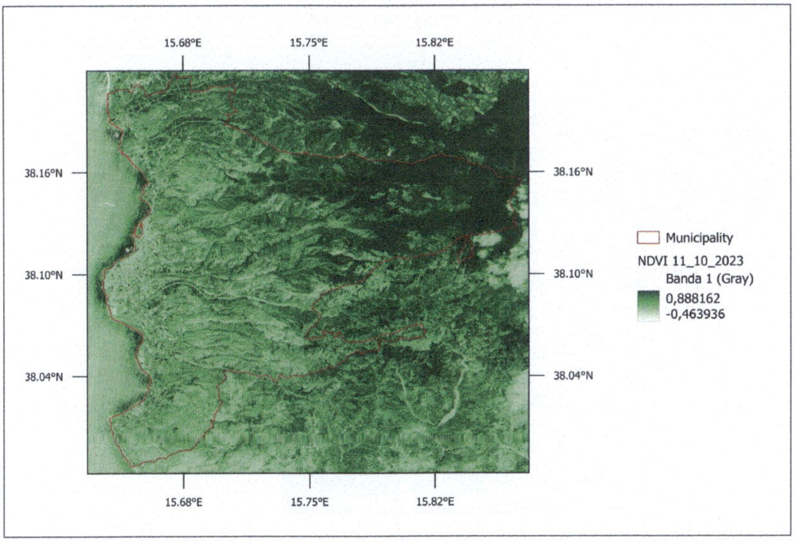

(b)

Fig. 5. Municipality of Reggio Calabria area average NDVI 2019 (a) and 2023 (b). Source: Author's elaboration

been merged with NDVI and LAI data by reporting their variation in the period 2019–2023 associated with each land use parcel. By associating these indicators with Land Use parcels, the study aimed to detect and interpret variations in urban environmental sustainability over the specified time frame (2019 to 2023). This comprehensive approach allowed a better understanding of how remote sensing technologies contribute to assessing UES and informed urban planning from a data-driven perspective.

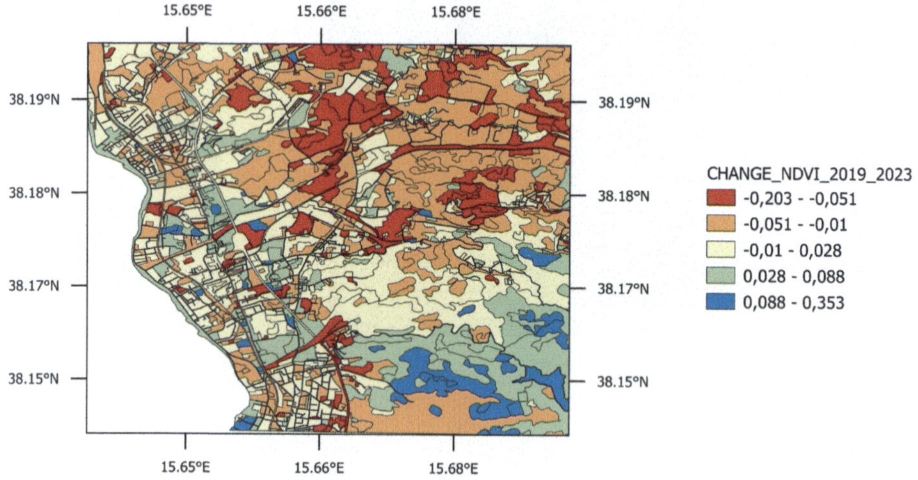

(a)

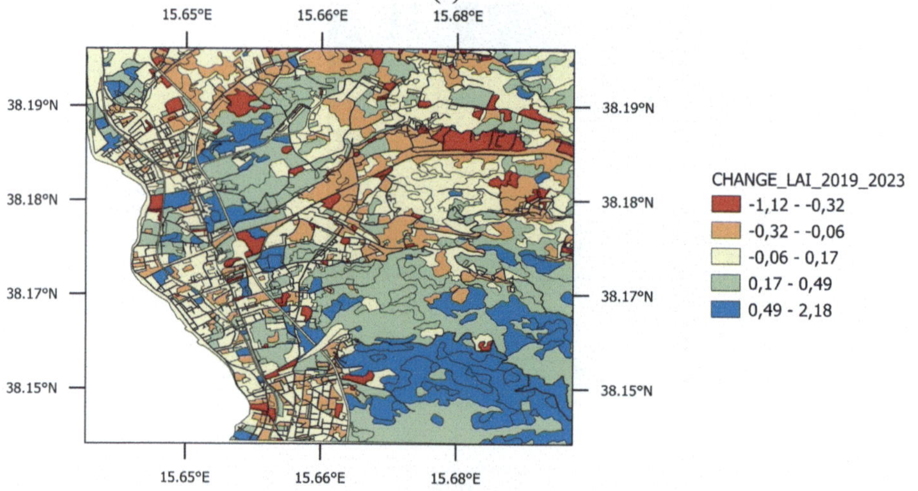

(b)

Fig. 6. Municipality of Reggio Calabria northern area (Area A) NDVI and LAI change (2019–2023) by Land Use. Source: Author's elaboration

The detection of NDVI and LAI through remote sensing technologies and its combination with other datasets through the urban informatics approach could result helpful in: identifying ESs supply areas; Prioritize areas for intervention in terms of ESs enhancement; provide useful information on the health of vegetation and natural ecosystems;

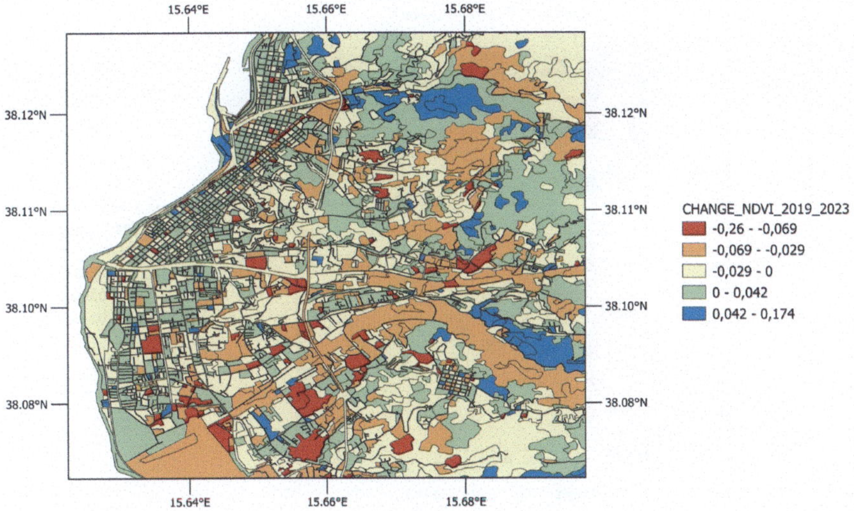

(a)

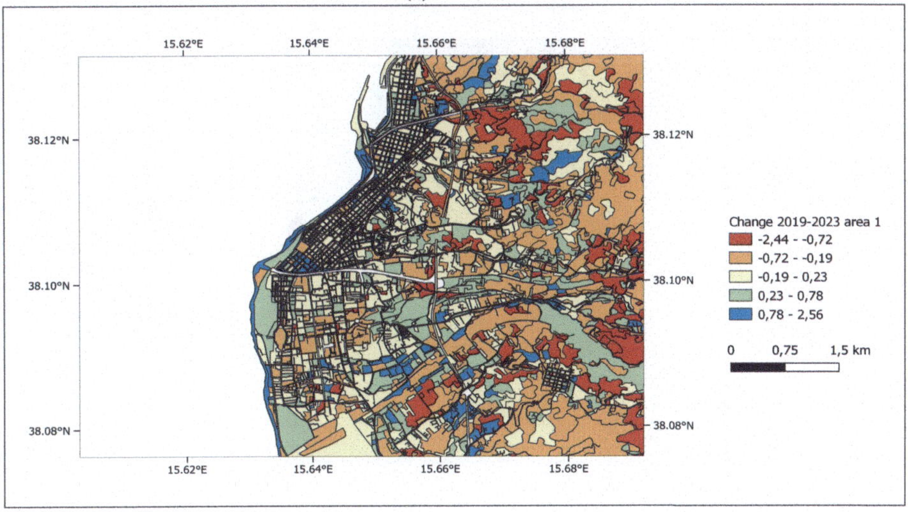

(b)

Fig. 7. Municipality of Reggio Calabria Urban Core (Area B) NDVI and LAI change (2019–2023) by Land Use. Source: Author's elaboration

identifying possible vegetation areas under (climate) stress factor or areas under maintenance/improvement; detecting possible area interested by fires; monitor the areas under transformations.

In summary, the results section presents a detailed account of the data analysis process, emphasizing the application of NDVI and LAI data as potentially useful in understanding and measuring changes in Urban Environmental Sustainability within the Reggio Calabria municipality.

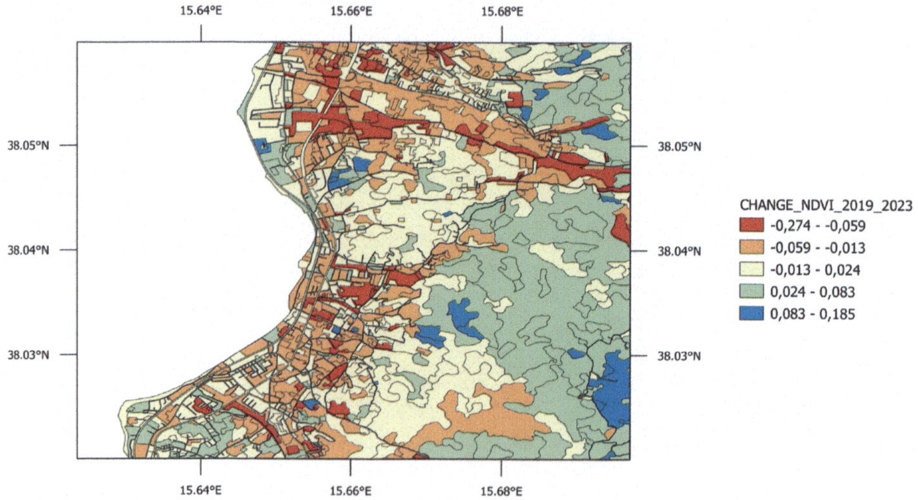

(a)

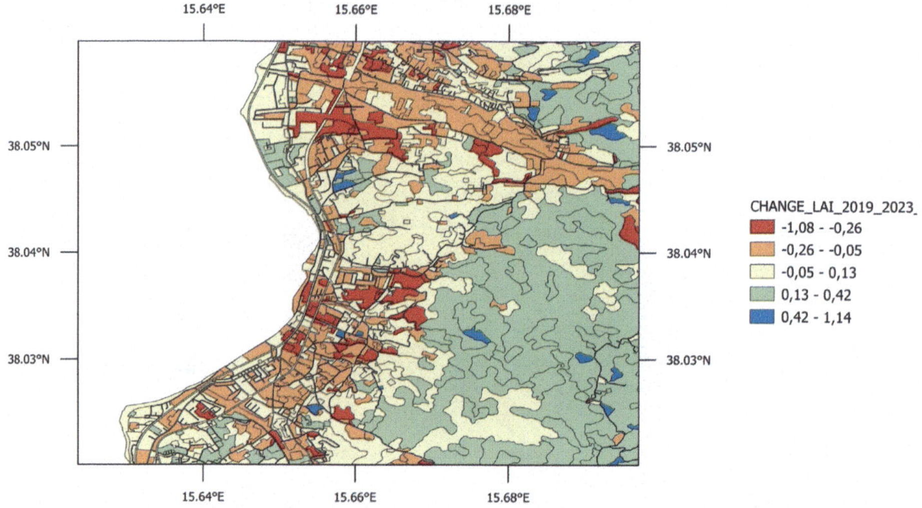

(b)

Fig. 8. Municipality of Reggio Calabria Southern (Area C) area NDVI and LAI change (2019–2023) by Land Use. Source: Author's elaboration.

5 Discussion and Conclusions

The paper explored data-driven planning approaches for measuring urban ecosystem services in cities by focusing on the potential of big data and urban informatics approaches. The literature review conducted has outlined how UES are pivotal in providing essential services to urban populations, encompassing clean air, water purification, climate regulation, and recreational spaces. Enhancing these services not only improves overall well-being but also boosts urban resilience. The integration of big data and urban informatics emerges as a potent approach for measuring and enhancing UES through urban planning. Leveraging diverse data sources, including satellite images and remote sensing data can facilitate data-driven decision-making and planning. In this direction, metrics like Vegetation Indexes (e.g., NDVI, LAI) play a crucial role in detecting UES in cities highlighting the transformative potential of data-driven planning in urban development. Such metrics have been detected in a test-case city capturing vegetation status in urban areas – one of the prerequisites for the spatial recognition of UES.

The study presents some limitations. The association between NDVI and LAI by Land Use parcel could present some biases. The areas where the high negative change in NDVI or LAI is detected could be both areas under urban transformation processes, in which the loss of vegetation could be due to the construction activities, or, in the case of the areas out of the urbanized core, it could indicate the occurrence of fires during the summer season. This element necessitates further investigation by examining the status of this area through official documentation and data. Further studies can widen the type of indicators and data to exploit for this purpose.

Despite its intrinsic limitation, the study presents some interesting implications for the urban planning dimension. Following Cortinovis and Geneletti [10], the integration of UES approaches into urban planning can support the prioritization of planning choices – at least in the initial phases of the planning process – and promote the definition of ES-enhancement objectives and targets orienting the implementation and monitoring processes. The advent of big data in cities, coupled with the emergence of urban informatics and data-driven urban planning, together with the development of transformative technologies, bring unprecedented opportunities for enhancing urban environments' efficiency, sustainability, and livability. Integrating big data allows for more informed decision-making, up-to-date responsiveness, and monitoring planning choices. As cities evolve into data-driven entities, urban planners could incorporate diverse forms of knowledge, embracing a holistic understanding of data and fostering inclusive decision-making processes in shaping the transition of cities. Urban informatics approaches allow the capture of ongoing transformative dynamics in cities, formulating new ideas on resilience, disruption, and climate change thanks to new data sources, also mainly open and public [19].

The contribution allows for outlining potential implications regarding data-driven city planning thanks to the multifunctional role played by big data and urban informatics. First, for the strategic planning dimension, the study suggests that big data and urban informatics can be crucial - by harnessing innovative technologies like remote sensing and utilizing big data – for strategically plan for the sustainable utilization of UES through the setting of long-term goals, identifying key areas for conservation, ensuring the equitable distribution of ecosystem services across the urban landscape, and catalyzing financial resources with specific goals for increasing resilience and sustainability. Second, embedding UES-related objectives and integrating UES data into urban planning has potential implications for comprehensive planning for supporting the development of inclusive and sustainable cities. Third, zoning regulations can benefit from the insights provided by UES data. The paper suggests that mapping indicators such as NDVI and LAI concerning land use categories can inform zoning regulations, which can be adapted to support critical ecosystem services, such as green spaces and areas with high UES values, by finding a balance between the demand and supply of UES in cities. This can contribute to creating more resilient and sustainable transformations urban zones. Fourth, From the urban governance perspective, the study emphasizes the importance of recognizing the impact of social, ecological, economic, and technological systems on the availability of UES. Urban governance needs to incorporate this understanding into decision-making processes. The rise of big data and urban informatics presents opportunities for effective governance by providing insights into people's interactions with urban nature and supporting the development of policies that address various levels of urban planning.

The relevance for cities, as for the test-case study area, the city of Reggio Calabria, follows the potential implications related to the urban planning and governance dimension. Specifically, such approaches can result helpful in detecting those areas in the city where UES are improving citizens quality of life and the same time those where living conditions are poorer, contributing to the prioritization of choices in terms for interventions. The detection of NDVI and LAI changes contribute also to the monitoring of urban transformations. Moreover, such areas can result useful for catalyzing the investments related to public programs by maximizing the effects for the transition of cities towards sustainability.

Several promising research directions could be explored in the context of UES and the integration of data-driven approaches. First, in terms of UES Mapping and Assessment, as big data analytics can support the analysis of temporal dynamics of UES for the examination of climate patterns and long-term trends, which is useful also for the definition of predictive models and facilitates multi-scale analysis for capturing the interconnectedness of UES across different spatial levels. Second, in terms of integrating social and ecological data, by examining ways to integrate social and ecological data to understand the complex interactions between human activities and urban ecosystems and inform decision-making for sustainable urban development.

Exploring these research trajectories can contribute to a more comprehensive understanding of the UES dynamics in cities by providing interesting insights for data-driven urban planning strategies that prioritize their transition toward sustainability.

Acknowledgements. This research work is the result of the synergistic activities related to the following PNRR Research Projects funded by the European Union – NextGenerationEU: "Pilot Project 4.6.1 - Climate adaptation plans for the reduction of the ecological footprint and ecological debt, aimed at improving the conservation and transformation response in terms of resilience and quality of life in the Calabrian and Lucanian urban and territorial systems" project (Goal 4.6 - Planning for Climate Change to boost cultural and natural heritage: demand-oriented ecosystem services based on enabling ICT and AI technologies - Tech4You Innovation Ecosystem), Finanziato dall'Unione europea- Next Generation EU, Missione 4 Componente 2 - CUP C33C22000290006 (ECS_00000009); the "ECO- SET - A Multidisciplinary approach to plan Ecosystem Services for cities in Transition" project – Finanziato dall'Unione europea- Next Generation EU, Missione 4 Componente 2 - CUP B53D23029450001; the "PLANET – PLANning Ecosystem services for cities in Transition" project - Finanziato dall'Unione europea- Next Generation EU, Missione 4 Componente 2 - CUP C33C22000840006.

References

1. Bolund, P., Hunhammar, S.: Ecosystem Services in Urban Areas (1999)
2. Engin, Z., et al.: Data-driven urban management: mapping the landscape. J. Urban Manage. **9**(2), 140–150 (2020). https://doi.org/10.1016/j.jum.2019.12.001
3. Shen, J., Wang, Y.: Allocating and mapping ecosystem service demands with spatial flow from built-up areas to natural spaces. Science of the Total Environ. **798** (2021). https://doi.org/10.1016/j.scitotenv.2021.149330
4. Yu, D., Fang, C.: Urban remote sensing with spatial big data: a review and renewed perspective of urban studies in recent decades. Remote Sens (Basel) **15**(5), 1–34 (2023). https://doi.org/10.3390/rs15051307
5. McPhearson, T., Andersson, E., Elmqvist, T., Frantzeskaki, N.: Resilience of and through urban ecosystem services. Ecosyst. Serv. **12**, 152–156 (2015). https://doi.org/10.1016/j.ecoser.2014.07.012
6. Kremer, P., et al.: Key insights for the future of urban ecosystem services research. Ecology and Society **21**(2) (2016). https://doi.org/10.5751/ES-08445-210229
7. Tan, P.Y., et al.: A Conceptual Framework to Untangle the Concept of Urban Ecosystem Services. Elsevier B.V. (2020). https://doi.org/10.1016/j.landurbplan.2020.103837
8. United Nations, "System of Environmental-Economic Accounting-Ecosystem Accounting White cover (pre-edited) version (2021). https://seea.un.org/ecosystem-accounting
9. Scowen, M., Athanasiadis, I.N., Bullock, J.M., Eigenbrod, F., Willcock, S.: The current and future uses of machine learning in ecosystem service research. Science of the Total Environ. **799** (2021). https://doi.org/10.1016/j.scitotenv.2021.149263
10. Cortinovis, C., Geneletti, D.: Ecosystem services in urban plans: what is there, and what is still needed for better decisions. Land Use Policy **70**, 298–312 (2018). https://doi.org/10.1016/j.landusepol.2017.10.017
11. Alberti, M., McPhearson, T., Gonzalez, A.: Embracing urban complexity. In: Urban Planet, Cambridge University Press, pp. 45–67 (2018). https://doi.org/10.1017/9781316647554.004
12. Kitchin, R., Lauriault, T.P., McArdle, G.: Data and the City. Routledge (2017). www.routledge.com/
13. O'Brien, D., Informatics, U.: Using Big Data to Understand and Serve Communities. CRC Press, Boca Raton, FL, Chapman Hall (2022)
14. Thakuriah, P., Tilahun, N., Zellner, M.: Big data and urban informatics: innovations and challenges to urban planning and knowledge discovery. In: Seeing Cities Through Big Data. Springer Geography, Thakuriah, P., Tilahun, N., Zellner, M. (eds.) Springer Geography. Springer, Cham, pp. 11–45 (2017). https://doi.org/10.1007/978-3-319-40902-3_2
15. Kitchin, R.: Big Data, new epistemologies and paradigm shifts. Big Data Soc. **1**(1) (2014). https://doi.org/10.1177/2053951714528481
16. Kitchin, R., McArdle, G.: What makes big data, big data? exploring the ontological characteristics of 26 datasets. Big Data Soc. **3**(1), 1 (2016). https://doi.org/10.1177/2053951716631130
17. Pan, H., Geertman, S., Deal, B.: What does Urban Informatics Add to Planning Support Technology? SAGE Publications Ltd. (2020) https://doi.org/10.1177/2399808320945453
18. Kontokosta, C.E.: Urban informatics in the science and practice of planning. J. Plan. Educ. Res. **41**(4), 382–395 (2021). https://doi.org/10.1177/0739456X18793716
19. Batty, M.: Urban Informatics and Big Data A Report to the ESRC Cities Expert Group 1 Preamble and Summary (2013)
20. Manley, K., Nyelele, C., Egoh, B.N.: A Review of Machine Learning and Big Data Applications in Addressing Ecosystem Service Research Gaps. Elsevier B.V. (2022). https://doi.org/10.1016/j.ecoser.2022.101478

21. Vysna, V., et al.: Accounting for ecosystems and their services in the European Union (INCA). Final report from phase II of the INCA project aiming to develop a pilot for an integrated system of ecosystem accounts for the EU.," Luxembourg (2021)
22. Diran, D., Hoppe, T., Ubacht, J., Slob, A., Blok, K.: A data ecosystem for data-driven thermal energy transition: reflection on current practice and suggestions for re-design. Energies (Basel) **13**(2) (2020). https://doi.org/10.3390/en13020444
23. Kabisch, N., Stadler, J., Korn, H., Bonn, A.: Nature-Based Solutions to Climate Change Mitigation and Adaptation in Urban Areas (2016)
24. Van Bodegom, P., et al.: Issue Paper New Technologies and Approaches to Measure Ecosystem Services Locally and to Engage Local Stakeholders (2020)
25. Ahern, J., Cilliers, S., Niemelä, J.: The concept of ecosystem services in adaptive urban planning and design: a framework for supporting innovation. Landsc. Urban Plan. **125**, 254–259 (2014). https://doi.org/10.1016/j.landurbplan.2014.01.020
26. Gómez-Baggethun, E., Barton, D.N.: Classifying and valuing ecosystem services for urban planning. Ecol. Econ. **86**, 235–245 (2013). https://doi.org/10.1016/j.ecolecon.2012.08.019
27. Zhao, J., et al.: A land cover classification method for high-resolution remote sensing images based on ndvi deep learning fusion network. Remote Sens (Basel) **14**(21) (2022). https://doi.org/10.3390/rs14215455
28. Wang, Q., Adiku, S., Tenhunen, J., Granier, A.: On the relationship of NDVI with leaf area index in a deciduous forest site. Remote Sens. Environ. **94**(2), 244–255 (2005). https://doi.org/10.1016/j.rse.2004.10.006
29. Kollányi, L., Máté, K., Prohászka, V., Fóris, E.D., Sallay, Á.: Greenness indicator for spatial and settlement planning based on greenness indicator for spatial and settlement planning based on NDVI and LAI indicators NDVI and LAI Indicators **6** (2019). https://doi.org/10.7275/4wgj-t211
30. Ogawa, S., Saito, K., Aihara, M., Otowa, K.: Estimates of Lai for Forest Management in Okutama (2001). https://www.researchgate.net/publication/237427963

New Gentrification Patterns in the Context of Urban Transition. The Unequal Spatial Distribution of Green-Certified Buildings in the City of Boston

Carmelina Bevilacqua[1], Armando Cepeda-Guedea[2](✉), and Ugo Lombardo[3]

[1] Sapienza Università di Roma, 00196 Rome, RM, Italy
carmelina.bevilacqua@uniroma1.it
[2] Università degli Studi Mediterranea, 89124 Reggio Calabria, RC, Italy
armando.cepeda@unirc.it
[3] Università degli Studi di Palermo, 90133 Palermo, PA, Italy

Abstract. The building sector has a key role in the urban transition since changing from a conventional building stock to a green one is critical for the reduction of emissions. Currently, there is a noticeable difference in costs between conventional and green buildings that is eventually reflected in the final price, this gap can be interpreted as the costs of urban transition in the building sector. This difference in costs can be assumed either by the private sector, the government, or a combination of both. This article explores the most important factors that define the distribution of green-certified buildings as a proxy for urban transition at the neighbourhood level in a context where most green buildings are developed by private investors and where these practices are mainly concentrated in the centric and wealthiest neighbourhoods. To achieve this, the paper uses a case study of the city of Boston, where there is a high density of green-certified buildings. The article proves the pertinence of using LEED-certified buildings as a proxy to estimate the costs of urban transition and how their variables influence the final price, identifies the main factors that define the distribution of LEED-certified buildings in Boston's neighbourhoods and finally presents a first approximation to estimate the risk of green gentrification at the neighbourhood level. The paper expects to provide arguments to policymakers to invest and stimulate the presence of green-certified public buildings and housing in vulnerable neighbourhoods to achieve a more egalitarian urban transition and reduce the risk of green gentrification.

Keywords: Green Gentrification · Costs of Transition · Green Certified Buildings

1 Introduction

The building sector is the largest contributor to greenhouse gas emissions, representing roughly 30% of global emissions (Si-Wei and Gou, 2023). In the case of developed countries, cities such as Boston can reach 71% of the total emissions (Global Covenant

F. Moraci et al. (Eds.): NMP 2024, LNNS 1272, pp. 35–61, 2025.
https://doi.org/10.1007/978-3-031-82927-7_3

of Mayors, 2023). Reducing the emissions of the building sector plays a key role in sustainable development and achieving climate neutrality (Fastenrath and Braun 2018). Transitioning and upgrading to green buildings has become the dominant trend in recent times in the sector, supported by both soft and hard incentives such as subsidies and regulations (Si-Wei and Gou, 2023). A green building can be defined as the practice of creating structures and using processes that are environmentally responsible by using less energy and water, as well as using materials with less environmental impact and regenerative potential (Wen et al. 2020) (Liu et al. 2022). The growth of green buildings in cities is part of the "urban transition agenda", which is a global strategy aimed at addressing the global environmental crisis and moving towards climate neutrality (Affolderbach and Schulz 2018).

It is estimated that green buildings can decrease operational costs by up to 37%, have up to 23% higher occupancy rates, have higher rental income of up to 8%, and have faster sale times (Sokolowski, Maheshwari, and Malik 2019). As a result, there is a growing demand for green buildings worldwide (Tsai 2022). To address this trend, green building rating systems have emerged, aiming to provide standardized assessment and certification for various types of constructions. Currently, around 600 frameworks are used to evaluate and certify green buildings around the world (Si-Wei and Gou, 2023). While BREEAM (Building Research Establishment Environmental Assessment Method) and LEED (Leadership in Energy and Environmental Design) maintain prominent positions as widely recognized certification schemes worldwide (Si-Wei and Gou, 2023). BREEAM certification currently has 535,000 certified buildings and its presence is predominant in the European continent (BreGroup 2023). On the other hand, LEED certification has more than 100,000 certified buildings and is the most popular certification framework in the United States (U.S. Green Building Council 2022).

Green-certified building features have been proven to be aligned with the efforts to achieve climate neutrality and can be a reference to understanding the urban transition processes in cities (Affolderbach and Schulz 2018). Firstly, there is a clear synergy between green-certified buildings such as LEED and the Sustainable Development Goals (SDGs), mainly related to sustainable consumption and production (Goubran et al. 2023). The upgrade and transition of the conventional building stock to a green a high performance one represents a significant cost that has to be assumed by the private developers, public authorities or a combination of both (Haldrup and Snällfot 2014). Studies suggest that green building cost premiums can reach 21% in comparison with the conventional ones, which are reflected in the final sales price (Dwaikat and Ali 2016). In the United States, LEED-certified buildings have a higher average market sales price per square foot of 21.4% over non-LEED buildings (Cushman and Wakefield 2021). This gap can be interpreted as the costs of urban transition in the building sector.

Since the development of green buildings is largely dependent on the private sector (Haldrup and Snällfot 2014), and the cost premiums are mainly assumed by real estate developers who transfer the costs to the real estate buyer, leading to an increase in the land values (Dwaikat and Ali 2016) in places with a high concentration of green certified buildings and other green amenities. This situation can contribute to the "green gentrification" in cities, which occurs when the installation of new, improved, or remediated urban green spaces attracts increased investment and developer attention, forcing residents who are unable to adapt to the increased costs of living to relocate to peripheral areas and excluding them of the benefits of green buildings and green spaces and infrastructures in general (Sax, Nesbitt, and Quinton 2022). This provokes a conflict between the ambitions of local governments and their targets for lower CO2 emissions, higher energy standards (energy efficiency) and advances towards "green cities" (Haldrup and Snällfot 2014) for all their residents and the needs of the market to allocate green buildings in neighbourhoods that can provide the highest profit.

Public financed green buildings play a key role in the fair distribution of this practice to make it accessible to all, regardless of their income and geographical location (Gupta 2015) (Peng and Bai 2021). However, investments in green building development in the public sector for the short, medium, and long term are hampered by the fact that there is often insufficient long-term real estate management in the public sector (Vlaams Energiebedrijf 2019). This provokes a slow and inefficient implementation of green buildings in the sector. Local governments, in collaboration with private developers, are responsible for the development of public green affordable housing (Gupta 2015). To avoid the increase in the final price of green-certified social housing and to remain affordable for vulnerable populations, local governments can provide incentives to assist developers in building sustainable housing, such as making it easier to obtain planning permission (Gupta 2015) (Peng and Bai 2021). In addition, financially incentivizing the greening of social housing along with local incentives will encourage developers to increase their participation in the sector (Gupta 2015). As an example, whereas some cities require developers to allocate a proportion of development to social housing, similarly green measures can be regulated, and the provision of green housing can be incentivized (Gupta 2015). Finally, economic and market-based instruments, such as tax exemptions and reductions, can be complementary to regulatory and control instruments (Tsai 2022). It is critical a bigger participation of the public sector in the development of green buildings to ensure a fair distribution of them as green buildings, infrastructures and spaces and avoid green gentrification processes.

Considering this background, the paper argues that when the costs of transition as reflected in green-certified buildings are assumed by the private sector can lead to an uneven distribution of them and increase the risk of green gentrification in vulnerable neighborhoods when new privately financed green-certified developments arrive in the area and provoke the eventual displacement of the original population who cannot afford the new housing and living prices. This situation calls for a favourable public policy framework aimed at stimulating the desired market behaviour and the strengthening of the presence of publicly financed green-certified buildings. The reviewed literature outlines the challenges and risks of exclusion cities face in their efforts to renew their conventional building stock into a greener and more high-performance one to contribute to the objective of achieving climate neutrality, and the relevance to quantify the costs of urban transition that cities in collaboration with private developers should assume. For this reason, it is important to provide tools and methodologies to support a better understanding of the phenomenon and help decision makers to act in consequence. The paper attempts to answer the following question: Which are the most important factors that define the distribution of green-certified buildings as a proxy for urban transition at the neighbourhood level in a context where most green buildings are developed by private investors?

Specifically, the paper seeks to understand where the green-certified buildings such as LEED and by association urban transition concentrate and how they are distributed in a mainly market-driven system and identify where public authorities can interfere to ensure that all citizens benefit from the benefits of greener and sustainable neighbourhoods and reduce the risk of green gentrification. Also, the paper seeks to prove the synergies of LEED-certified buildings and urban transition and the level of influence of their variables in the final price of buildings, to use LEED-certified buildings as a proxy to explain processes and to quantify the possible costs of urban transitions. For this purpose, the paper proposes a multidisciplinary analytical approach using the city of Boston as an exploratory case study. The data was obtained from sources from private and public organisations. The choice of the city of Boston as a case study is explained because of its high density of green-certified buildings, more specifically LEED-certified (503 by October 2023) of different typologies that allow an adequate data set to execute the required analyses and that majority of these buildings are privately owned and developed representing 95.5% of the total. Also, the diversity of the 23 neighbourhoods of the city allows a better understanding of the distribution of green-certified buildings by providing enough diversity in relevant factors such as socioeconomic composition and geographical distance.

The paper is structured as follows. The literature reviews the challenges cities may face in their efforts to renew their conventional building stock into a greener and more high-performance one, the risks of green gentrification in cities, LEED certification as a standard to define green buildings, and the relevance to quantifying the costs of urban transition that cities in collaboration with private developers should assume. The methodology section explains the proposed steps and materials the paper follows to present their results. It also provides the context of the city of Boston and the presence of LEED-certified buildings in the city using maps and diagrams. The results section, supported by multiple regression and correlation analyses, first identifies the influence of LEED-certified variables in the final price and the weight of each of them. Subsequently, identify the main factors and their weights that define the distribution of LEED-certified buildings in the neighbourhoods of the city and finally identify the most vulnerable neighbourhoods to green gentrification through a first approach of a green gentrification risk index. The discussion and conclusion section argues the relevance of a favourable public policy framework aimed at stimulating the desired market behaviour for a fair distribution of green-certified buildings to avoid the negative effects of green gentrification and calls for the strengthening of the presence of publicly financed green-certified buildings to provide the benefits of greener neighbourhoods to all the inhabitant of the city disregarding their income.

2 Literature Review

2.1 The Challenges of Transforming the Conventional Building Stock into a Green One

It is widely accepted that transforming the conventional building stock into green and high-performance is the right path to achieve climate neutrality by 2050 (Tsai 2022) (European Commission 2019). Nevertheless, the implementation of green buildings in cities is not easy. The difficulty for cities to adopt faster green buildings can be explained because of five main barriers (Darko et al. 2017). The first barrier is the evident economic challenges presented as one of the biggest barriers to sustainable design and construction in the US and other countries (Simpeh et al. 2021). The construction cost of green buildings can be as high on average as 30 USD/m2 (Liu et al. 2022). In the case of LEED-certified buildings, there can be additional costs to achieve the certification requirements contributing to a possible low construction profitability and long payback periods (Uğur and Leblebici 2018). In developing countries such as China and India, there is a clear lack of demand for green buildings due to the elevated costs (Uğur and Leblebici 2018). The second barrier is related to the surrounding environment which can directly affect the full performance of green-certified buildings such as LEED affected by factors such as relative humidity, air temperature and velocity and average radiation temperature (Darko et al. 2017) (Çiner and Doğan-Sağlamtimur 2019). The third barrier is the technological challenges like the lack of widespread use of technologies, such as green roofs, climate efficient windows, solar shading devices, solar water heaters, and water treatment plants. Also, the use of new design tools for green building design such as Building Information Modeling (BIM) technology is not widespread (Chakravarthy et al. 2022). The fourth barrier is the institutional and policy challenges. National and local financial incentives are not sufficient to effectively motivate a bigger investment in green buildings. The fragmentation of legal and institutional frameworks hindered the development of green building development because of the lack of unified regulations and standards. Minimum legislation and penalties for non-compliance for the adoption of green practices in the building sector (Darko et al. 2017) (Gupta 2015) (Ayarkwa et al. 2022). The fifth and final barrier is the social-cultural challenges like the lack of awareness and overall sustainability knowledge and understanding for private developers and civil society of the long-term benefits of green buildings. Also, the ineffective collaboration among the different involved stakeholders in the green building scale-up process and the lack of public participation calling for a wider adoption of green buildings (Darko et al. 2017).

These barriers call for a better diffusion among decision-makers and civil society of the main benefits of adopting green buildings which are greater water and energy efficiency, and an improvement of the image and reputation of governments and the private sector. It suggested that the most important strategies to promote the adoption of green buildings are financial and better market-based incentives and better availability of information on the long-term benefits of green buildings.

2.2 An Approximation for the Costs of Urban Transition

Recent analysis suggests that low-carbon urban actions represent a USD 16.6 trillion economic opportunity worldwide (Floater et al. 2017). The value of the overall green building market is estimated to be USD 49 billion with an anticipated market value of USD 140 billion by 2013 (Tatari and Küçükvar 2011). Financing mitigation actions such as stimulating a bigger presence of green buildings in cities are critical to achieving the reduction of emissions and eventual climate neutrality pathway (Tsai 2022). In general, the implementation of green buildings comes at an additional cost that is likely to be incompatible with rational economic behaviour that seeks to maximise profits (Haldrup and Snällfot 2014). This creates a conflict with public authority's ambitions to create greener and more resilient cities in the context of the climate change emergency. Specifically, the objective of refurbish existing buildings and upgrade urban space to fit the "green city" ambition. Direct funding, such as grants and subsidies, can be a feasible financial instrument to enable authorities to conduct climate mitigation projects. Subsidies are provided in many countries to develop new green buildings or retrofit them following green design criteria to decarbonize existing building stock. Relevant studies have found that the design of subsidy is key to maximizing carbon reduction and achieving an equal distribution of green buildings and practices in cities that are largely dependent on the private sector since transitioning the building stock is to be financed by the market. This is relevant considering that the costs of the implementation of new policies to enforce the presence of such as a new green building code can increase the construction costs by 10.77% as the case in the state of California (Kim, Greene, and Kim 2014). This does not consider the extra cost that may imply having the building green-certified by relevant organizations like LEED or BREEAM.

The costs of green buildings can reach 21% representing a relevant factor for private and public developers to invest in this type of buildings (Dwaikat and Ali 2016). In the United States the most popular certification which is LEED the construction costs can increase from 7.43% to 9.43% (Cushman and Wakefield 2021). This difference grows considering that LEED-certified buildings have 21.4% higher average market sales (Cushman and Wakefield 2021). This can be interpreted first as the reputational premium that LEED represents and the value that the market and the final users are willing to pay to have, use and live in green-certified buildings.

2.3 The Risks of Green Gentrification

Interventions aimed to increase and improve green spaces and practices have been proven beneficial for the health and overall well-being of citizens (Cole et al. 2017). Despite this potential, however, studies have revealed inequities in the distribution of the benefits of urban green practices among residents (Shonkoff et al. 2011) (Rigolon and Németh 2019). Green gentrification occurs when the installation of new, improved, or remediated urban green spaces such as parks, water bodies and green buildings attracts increased investment and private developers' attention, eventually increasing the land value of a neighbourhood or an urban area (Sax, Nesbitt, and Quinton 2022). This situation forces residents who are unable to adapt to the increased costs of living to relocate to more affordable locations, excluding them from the benefits of green practices. In the absence of physical displacement, it's possible a psychological exclusion manifested as a sense of exclusion among residents, increasing barriers to green space access, or a change in the qualities and uses of these spaces (Sax, Nesbitt, and Quinton 2022).

As cities are increasingly looking to urban greening to achieve their sustainability and carbon reduction objectives. Concerns have emerged that green gentrification will become an increasingly relevant topic in urban planning (Zuniga-Teran et al. 2021). Without proper consideration by public authorities for the potential adverse socio-spatial changes elicited by urban greening, the displacement of historically vulnerable residents will remain a prominent threat (Anguelovski et al. 2018). Studies suggest some reasons why injustices exist around the distribution of green practices which include green buildings. First, socio-economic factors play a key role. Minority and low-income groups have disproportionately less access to green space and green practices (Sax, Nesbitt, and Quinton 2022). This phenomenon can be partially explained by historic practices associated with land use, zoning, and housing patterns. Low-income communities generally attract less public investment and face greater challenges associated with legacy pollution (Taylor 2011). On the policy-making side, the lack of equitable investment across the city is reflected in the zoning, land use, and housing plans that fail to provide the necessary urban green infrastructure to achieve equity in the distribution of green practices (Ferguson et al. 2018). This situation calls for public authorities to consider and assess their unbalanced power structures, democratize their decision-making processes regarding the implementation of green practices, and leverage the benefits on green practices investments.

2.4 LEED Certification

The Leadership in Energy and Environmental Design (LEED) certification stimulates sustainable building design that includes reductions in energy, water, and building materials consumption, while at the same time enhancing occupant health and overall community connectivity (USGBC 2022). LEED certification is one of the most common rating systems for green buildings worldwide (Mady 2017). It provides a framework that real estate developers can apply to create healthy, highly efficient, and cost-saving green buildings. LEED certification is a recognized symbol of sustainability achievement all over the world with 100,000 certified buildings by 2022 (USGBC 2022). LEED certification story began when its first version was released in the year 2000 by the United States Green Building Council (USGBC) which develops and administrates the certification system (Champagne and Atkas 2016). The USGBC was established in 1993 to promote sustainability in the building's design and construction processes. It was found that LEED-certified buildings produced a significantly lower concentration of greenhouse gases: 50% less due to water consumption, 48% less due to solid waste management, and 5% less due to transportation compared to baseline non-certified buildings (Pushkar 2021). This proves the relevance of LEED-certified buildings in the urban transition processes.

The certification is constantly revised and updated, its latest version being LEED v4 launched in 2013. Nevertheless, currently, the most widely adopted version is LEED v3 (Amiri, Ottelin, and Sorvari 2019) launched in 2009 and operative until 2022 (USGBC

2022), which includes a collection of rating systems for design, construction, and operation of different types of buildings. In the certification, the allocation of points between credits is based on the potential environmental impacts and human benefits of each credit concerning a set of impact categories. The resulting allocation of points among credits is called credit weighting. LEED v3 2009 is composed of five main categories: sustainable sites (26 points), water efficiency (14 points), energy and atmosphere (35 points), materials and resources (10 points), and indoor environmental quality (15 points). The scoreboard considers two additional bonus categories: innovation in design (ID, 6 points) and regional priority (RP,4 points) giving a possible total of 110 points (USGBC 2010). It is important to consider that the regional priority category in the GBIG portal is merged into the other categories, so it is not directly visible in the score like the Innovation and Design bonus category. The 10 bonus credits aim to stimulate environmentally sensitive projects to work towards higher achievements (Pushkar 2021). The certification level depends on the points allocated based on how well a building fulfils the specific criteria in the different assessment categories and includes the following: Certified (40–49 points), Silver (50–59 points), Gold (60–79 points), and Platinum (80 + points) (Amiri et al. 2019).

3 Methodology

The research methodology explores the most important factors that define the distribution of green-certified (LEED) buildings as a proxy for urban transition at the neighbourhood level in a context where most green buildings are developed by private investors, and where these practices are mainly concentrated in the more centric and wealthiest neighbourhoods of the city of Boston. This is to subsequently obtain a first approach for a green gentrification risk index for the city at the neighbourhood level (see Fig. 1). Also, it proves the influence of the LEED certification variables (v2009) in the final price of the buildings, which can represent a valid approximation for the costs of urban transition in Boston. It intends to provide decision-makers tools to invest and stimulate the presence of green-certified public buildings and housing in vulnerable neighbourhoods to achieve a more egalitarian urban transition and reduce the risk of green gentrification. The methodology is based on a single exploratory case study (Priya 2020) using secondary data obtained from the City of Boston data platform (City of Boston 2023), the Boston Planning and Development Agency reports (BDPA 2023), the U.S. Census data portal (US Census 2023), Boston Housing Authority data (BHA 2023), the U.S. Green Building Council data (USGBC 2023), Green Building Information Gateway (GBIG 2023) and the real estate websites Zillow (Zillow 2023) and Commercial Café (Commercial Café 2023). The authors use the case study approach which allows in-depth, multi-faceted explorations of complex issues in their real-life settings. The reason to choose the city of Boston as a case study is because of its high density of green-certified buildings, specifically 503 LEED-certified buildings by October 2023 (USGBC 2023), the commitment of the city of Boston government to transition to a more sustainable and resilient city as reflected in the Boston 2030 (City of Boston 2018) strategy, the Boston Climate Action Plan (City of Boston 2019) and the predominant private ownership and development of green buildings, only 6.5% of the total LEED-certified buildings in the city are developed or partially developed by public funds. The methodological approach for the paper is articulated as follows.

1. Provide a context of the city of Boston, their neighbourhoods, the policy regarding the green building development and the LEED-certified buildings distribution and their evolution from the period (2015–2023).
2. Estimate the influence of LEED-certified variables in the final price in the city of Boston and prove if they are pertinent as a proxy for urban transition costs. To do this the authors selected 40 buildings from the LEED v3 2009 certification as sample. The selection of the v3 2009 version is because it is the most numerous in the city (367 of the total 503 buildings) (USGBC 2023) and was valid until 2022, from this version it was selected two representative categories, 20 from the Existing Buildings category (64 total) and 20 from the New Construction category (80 total). For each category, a model will be run. The selected method is a multiple regression model using RStudio software. The dependent variable is the market price of the selected LEED-certified buildings in 2023 per sqft. Instead, the explanatory variables considered as the transition costs are the six main categories of LEED v3 2009 certification criteria used. The initial hypothesis (null hypothesis) that the authors want to prove is that the price level can be determined by the following variables:

$$Y_p = \beta_0 + \beta_1 l_1 + \beta_2 l_2 + \beta_3 l_3 + \beta_4 l_4 + \beta_5 l_5 + \beta_6 l_6 + u_i$$

   where:

   l_1 = Energy and Atmosphere;
   l_2 = Materials and resources;
   l_3 = Indoor environmental quality;
   l_4 = Sustainable sites;
   l_5 = Water efficiency;
   l_6 = Innovation in design;

   The statistical significance threshold established for the model is with a p-value at least less than 0.1, which provides at least moderate evidence against the null hypothesis. To obtain the importance of each variable, the authors used the Random Forest method, which was supported again with R software. This intends to obtain a closer approach to the real weights of variables to the final price. The results of these models are then reinforced with a correlation model to estimate if there is a correlation between the presence of LEED buildings and the residential market value in the 23 neighbourhoods of Boston.
3. Once the pertinence to use the LEED-certified buildings as a proxy for urban transition costs is established. The next step is explaining the main factors that determine the distribution of LEED-certified buildings and, eventually, urban transition in the neighbourhoods in the city of Boston. The selected method is a stepwise multiple regression, which allows the adaptation of regression models in which the choice of predictive variables is made using an automatic procedure (the paper used R software). Specifically, it was decided to start from a model that considered an initial set of variables and then adapt it through successive steps based on the results obtained. The dependent variable is the number of LEED-certified buildings in Boston by 2021

(because reliable socioeconomic data arrives until 2021) in each of the 23 neighbourhoods of Boston. Instead, the explanatory variables considered are the seven main factors that led to gentrification in the United States (National Neighborhoods Indicators Partnership 2017) (National Community Reinvestment Coalition 2020). The initial hypothesis (null hypothesis) that the authors want to prove is that the price level can be determined by the following variables:

$$Y_b = \beta_0 + \beta_1 f_1 + \beta_2 f_2 + \beta_3 f_3 + \beta_4 f_4 + \beta_5 f_5 + \beta_6 f_6 + \beta_7 f_7 + u_i$$

where:

f_1 = Distance to downtown (km);
f_2 = White population (non-Hispanic white) (%);
f_3 = Population with bachelor's degree (%);
f_4 = Residential market value;
f_5 = Renters (%);
f_6 = Residents below poverty line (%);

The statistical significance threshold established for the model is with a p-value at least less than 0.1 (typically ≤ 0.1), which provides at least moderate evidence against the null hypothesis. To obtain the importance of each variable, the authors used the Random Forest method, which was supported again with R software. This intends to obtain a closer to obtain a closer approach to the real weights of variables to the final price.

4. Once the main five factors were identified, all the values were standardized on a scale of 1 to 5 and then multiplied with the obtained weight in the Random Forest model. This is to obtain a final score per each neighbourhood again on a scale of 1 to 5. This is to obtain a first approach to a green gentrification risk index. Where 5 is the lowest risk of being green gentrified and 1 is the highest risk of green gentrification. The identified neighbourhoods with the highest risk of green gentrification, according to the authors, should be the ones where public authorities should invest and incentivise more the presence of green-certified buildings to avoid possible green gentrification in vulnerable neighbourhoods and achieve a fairer urban transition.

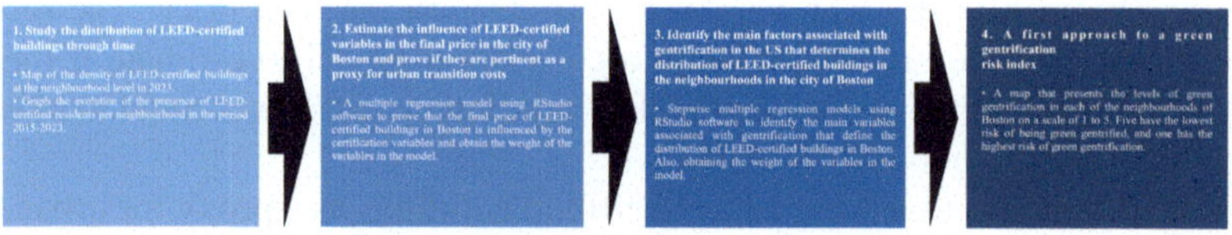

Fig. 1. Diagram of the methodologic flow of the article

3.1 City of Boston Green Buildings Distribution

The city of Boston since 2016 is part of the Global Covenant of Mayors for Climate & Energy and has committed to support action to combat climate change and work towards a resilient and low-emission society (Global Covenant of Mayors 2023). Boston was recognized as an A-List by CDP in 2022 (CDP 2022) by receiving a high score for their transparency and bold climate action and their achievements, vision, and commitment to the fight against climate change. This is reflected in their Climate Action Plan, updated in 2019 and current until 2024, which considers the following measures and is aligned with the broader initiative Imagine Boston 2030, which is the first citywide plan in more than 50 years that visualizes a city of Boston with quality jobs, a competitive economy, affordable housing, and better prepared for climate change.

Boston is a global city located on the northeastern coast of the United States. According to the US Census (2023), the city has a population of 650,706, while its metropolitan area, as estimated by the Metropolitan Area Planning Council (2023), is home to approximately 3,394,090 residents. Boston is officially divided into 24 neighborhoods, as recognized by the City of Boston government (2023), which serve as distinctive elements of pride and identity for its residents. (Note: Harbor Island is not included in this study). Regarding income using data from the BDPA in the period 2015–2021 (BDPA 2023), the authors produced a map and a graph showing the levels of income of the neighbourhoods of the City of Boston (see Fig. 2). Roxbury, in the western part of the city, is the neighbourhood with the lowest income, with an average of 37,844 USD per year. The southern part of Boston has a low-medium to high-medium income, with a representative neighbourhood such as Dorchester having an average income of 61,468.00 USD, below the city's average of 81,744 USD. The wealthiest parts of the city are located in the centre and north, where the main corporate and academic activity of the city is located, with the South Boston Waterfront being the wealthiest one with 167,446 USD per year because of the recent luxury residential development built in the area.

The spatial distribution of the 503 LEED-certified buildings in Boston, regardless of the typology supported by data from the USGBC, is uneven because most of them are in the downtown area of the city where the wealthiest neighbourhoods are located (see Fig. 3). This is evident in the city's urban landscape, where the most modern buildings are in the wealthiest areas of the city. Meanwhile, the traditional building stock is still predominant in peripheral neighbourhoods (see Fig. 4). This can be explained by the fact that most of the certified buildings are developed by private developers (95.5% of the total) who are expecting revenue by locating them in areas where people can afford to pay the extra value that implies a LEED-certified building (GBIG 2023). In the vulnerable areas of the city in the western and southern parts, the few LEED buildings are mainly from the public (6.5% of the total) and retail sectors because these areas are not as attractive as the central and wealthiest areas of the city. The current distribution of the certified LEED buildings in the city of Boston only allows for statistical study of the downtown area of the city because, in the peripheral areas, there are not enough buildings to study. The uneven distribution of LEED certified can be considered a challenge in the transition to more sustainable and resilient cities and the evolution from sustainable to positive buildings.

In the period 2015–2023, it can be seen a constant growth of the presence of LEED-Certified buildings in the city of Boston in consonance with the objectives of the city to have a greener and more sustainable city (see Fig. 5). Nevertheless, this growth is

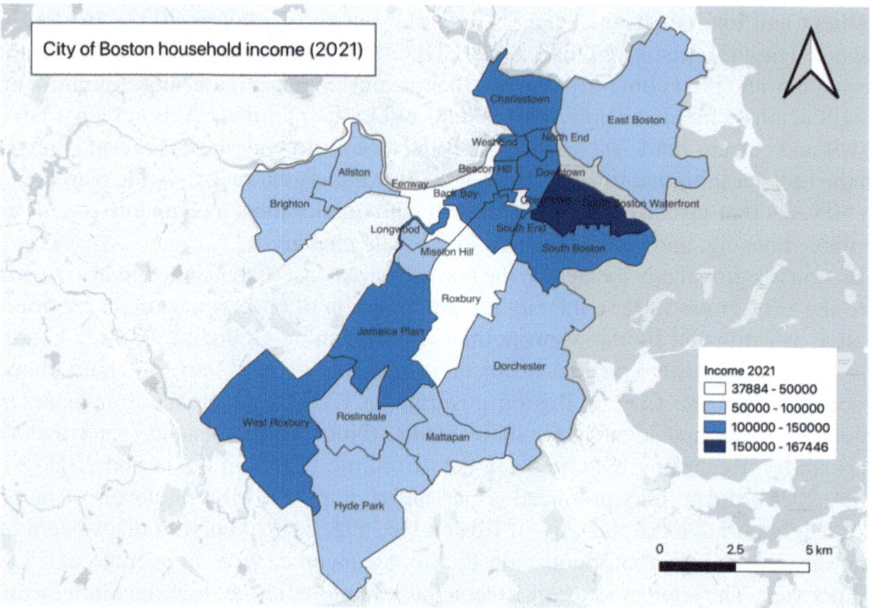

Fig. 2. Boston neighbourhoods and the evolution of the household income per neighbourhood in 2021

not fairly distributed among the neighbourhoods of the city. In 2023, Downtown is the neighbourhood with the highest density of LEED-certified buildings, with 151 (30% of the total) being a considerable part of the total. In second place comes South Boston Waterfront, which has 67 (13.3% of the total), being less than half of downtown. Finally, in third place comes Back Bay, which has 40 buildings (7.9% of the total). This means that more than half of the LEED-Certified buildings (51.2%) are concentrated in just 3 of the 24 neighbourhoods of the city. Meanwhile, lower-income neighbourhoods such as Rosindale and Mattapan just have 1 LEED-Certified building each.

4 Results

4.1 The Influence of LEED Certification Variables in the Final Price in Boston

The data consists of 40 observations from the LEED v3 2009 certification 20 from the Existing Buildings category and 20 from the New Construction category, intending to do one model per each category (see Fig. 6). The selected method is a multiple regression model using RStudio software. The dependent variable is the market price of the selected LEED-certified buildings in 2023 per sqft. Instead, the explanatory variables considered the transition costs are the six main categories LEED v3 2009 certification criteria used. The initial hypothesis (null hypothesis) that the authors want to prove is that the price level can be determined by the following variables:

$$Y_p = \beta_0 + \beta_1 l_1 + \beta_2 l_2 + \beta_3 l_3 + \beta_4 l_4 + \beta_5 l_5 + \beta_6 l_6 + u_i$$

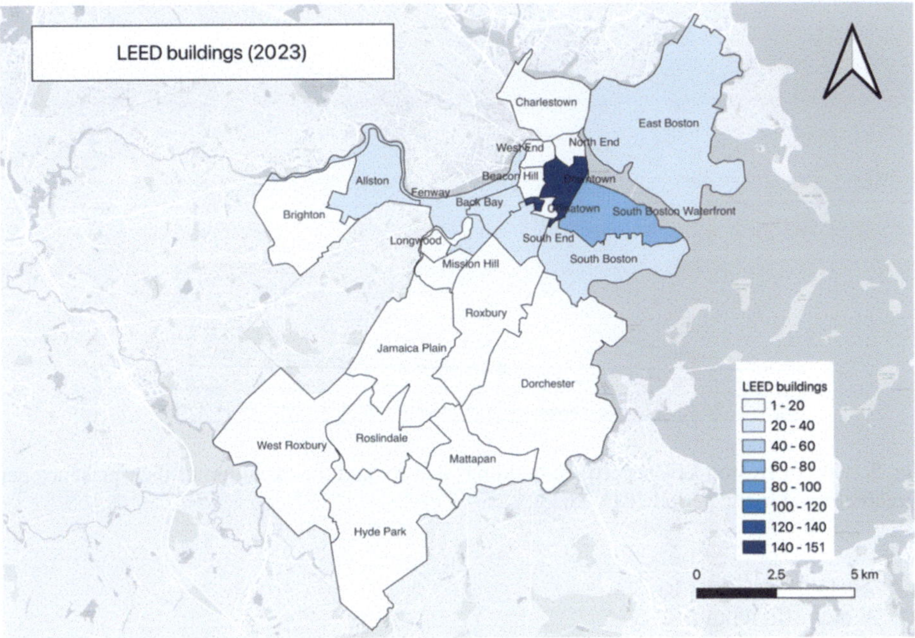

Fig. 3. Spatial distribution of the LEED-certified buildings in the city of Boston by density

Fig. 4. Examples of the building stock in Boston. A) Conventional buildings of the XX century in Fenway neighbourhood. B) The Hub on Causeway in the North End neighbourhood, one of the LEED-certified buildings with the highest score (85 points).

where:

l_1 = Energy and Atmosphere;
l_2 = Materials and resources;
l_3 = Indoor environmental quality;

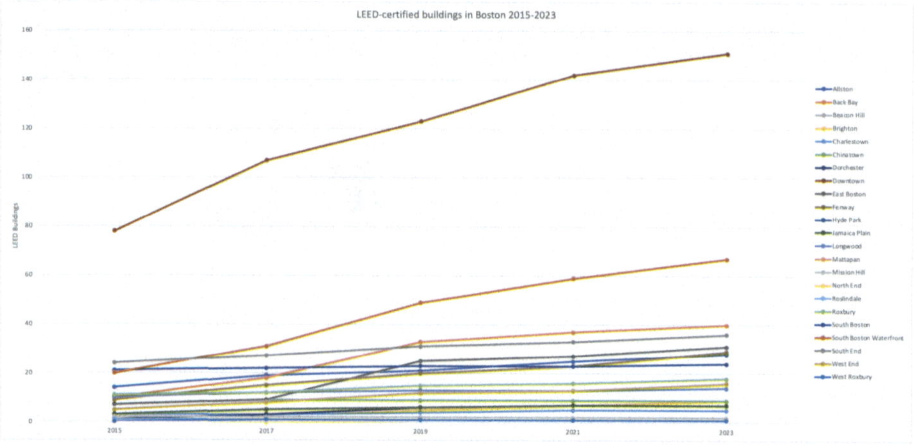

Fig. 5. City of Boston LEED certified buildings density and the evolution of their presence per neighbourhood in the period 2015–2023

l_4 = Sustainable sites;
l_5 = Water efficiency;
l_6 = Innovation in design;

The statically significance threshold established for the model is with a p-value at least less than 0.1 (typically ≤ 0.1) that provides at least moderate evidence against the null hypothesis. To obtain the importance of each variable, the authors use the Random Forest method supported again with R software).

The first multiple regression model from the LEED v3 2009 from the Existing Buildings category, which includes the six variables for the LEED v3 EB 2009 certification (see Table 2), shows an effective relationship between the variables and the price of the buildings. This is demonstrated by the R squared equal to 0.7375, for which 73.75% of the phenomena are explained by these variables, suggesting that there is a strong fit to the data in the model. Above all, the p-value is equal to 0.003192, which indicates a very strong level of significance. Therefore, the null hypothesis according to which these variables affect the level of building prices can be accepted since the p-value is above the significance level of ≤ 0.1 (Table 1).

Using the Random Forest package, it was possible to determine the importance of each LEED variable in the model. The results show that the most relevant variable is L5, which is water efficiency, representing 30.79%, closely followed by L3, which is Indoor environmental quality and represents 27.56%, and fairly close is L1, which is Energy and Atmosphere, with 23.50%. This means that three variables explain 81.85% of the model. The three variables with less influence in the model are L4, which is Sustainable sites with 14.03%; L2, which is Materials and resources with 2.78%; and L6, which is Innovation in Design, which has a lesser value with 1.34%.

The second multiple regression model from the LEED v3 2009 from the New Buildings category, which includes the six variables for the LEED v3 EB 2009 certification (see Table 2), shows an acceptable relationship between the variables and the price of

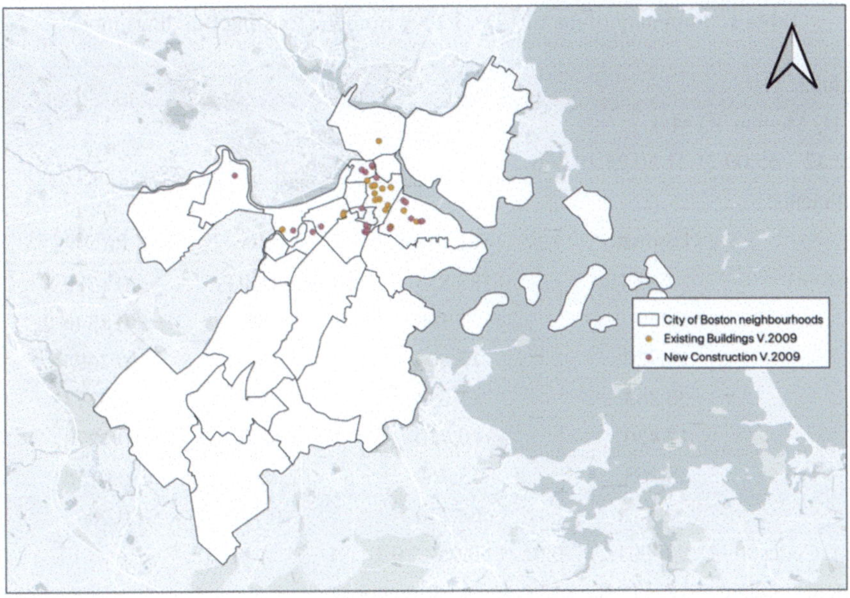

Fig. 6. Location of the selected LEED-certified buildings for the models

the buildings. This is demonstrated by the R squared equal to 0.5412, for which 54.12% of the phenomena are explained by these variables, suggesting that the model is a strong fit for the data. Above all, the p-value is equal to 0.04172, which indicates a strong level of significance. Therefore, the null hypothesis according to which these variables affect the level of building prices can be accepted since the p-value is above the significance level of ≤ 0.1.

Using the Random Forest package, it was possible to obtain the importance of each LEED variable for the second model. The results show again that the most relevant variable is L5, which is water efficiency, represents 23.47%, followed by L4, which is Sustainable Sites and represents 20.46%, and fairly close is L1, which is Energy and Atmosphere with 17.83%. This means that three variables explain 61.76% of the model. The three variables with less influence in the model are L3, which is Indoor environmental quality at 15.25%; L6, which is Innovation in Design at 12.17%; and finally, L2, which is Materials and resources, which has a lesser value at 10.80%. The second model shows a more equilibrated influence of the variables than the first one.

To reinforce the results, the authors consider it pertinent to run a correlation test to observe if the presence of LEED buildings is correlated with an increase in the price of buildings in Boston in the period 2015–2023 and see if there is an increase or decrease of this correlation across the time. The data was retrieved from the Zillow Home Value Index from 2015–2023, and the number of LEED-certified buildings registered at the USGBC site during the same time period (see Table 3).

The table shows that there is a strong and growing positive correlation between the presence of LEED buildings and the price of residential buildings. This test helps to understand better the influence of LEED-certified buildings in the real estate market in

Table 1. Summary of the LEED v3 2009 from the Existing Buildings model

Residuals:				
Min 1Q Median 3Q Max				
−222.42 −55.00 21.31 50.25 193.66				
Coefficients:				
	Estimate	Std. Error	t value	Pr(>|t|)
(Intercept)	−581.106	388.855	−1.494	0.1589
L1	1.340	7.169	0.187	0.8546
L2	15.920	13.804	1.153	0.2695
L3	36.263	15.634	2.319	0.0373
L4	17.420	10.710	1.627	0.1278
L5	38.263	13.764	2.780	0.0156
L6	−8.178	68.270	−0.120	0.9065
Signif. Codes: 0 '***' 0.001 '**' 0.01 '*' 0.05 '.' 0.1 ' ' 1				
Residual standard error: 118.8 on 13 degrees of freedom				
Multiple R-squared: 0.7375 Adjusted R-squared: 0.6164				
F-statistic: 6.088 on 6 and 13 DF, **p-value: 0.003192**				

the city of Boston. All the models presented in this section help to prove the adequateness of using LEED-certified buildings as a proxy for the costs of urban transition at least in the city of Boston. This is by proving that the LEED variables associated with urban transition have a real influence on the final price and that their presence is correlated with the increase in the price of buildings.

4.2 The Factors that Determine the Distribution of LEED-Certified Buildings in the City of Boston

Using as a starting point the variable that previous models establish to determine the risk of gentrification in Boston, the authors use a stepwise multiple regression to determine which of these variables explains the distribution of LEED-certified buildings and, by consequence, the possible risk of green gentrification. The dependent variable is the number of LEED-certified buildings in 2021 in each of the 23 neighbourhoods of Boston. Instead, the explanatory variables considered the seven main factors that lead to gentrification in the United States. The initial hypothesis (null hypothesis) that the authors want to prove is that the price level can be determined by the following variables:

$$Y_b = \beta_0 + \beta_1 f_1 + \beta_2 f_2 + \beta_3 f_3 + \beta_4 f_4 + \beta_5 f_5 + \beta_6 f_6 + \beta_7 f_7 + u_i$$

where:

f_1 = Distance to downtown (km);

Table 2. Summary of the LEED v3 2009 from the New Buildings model

Residuals:				
Min 1Q Median 3Q Max				
−768.1 −382.0 −154.1 419.1 1332.8				
Coefficients:				
	Estimate Std	Error	t value	Pr(>|t|)
(Intercept)	7579.778	1980.402	3.827	0.00165 **
L1	−34.585	35.013	−0.988	0.33893
L2	34.304	83.031	0.413	0.68535
L3	−231.738	115.042	−2.014	0.06226
L4	−90.947	49.539	−1.836	0.08628
L5	−424.479	112.642	−3.768	0.00186 **
L6	−5.959	148.147	−0.040	0.96844
Signif. Codes: 0 '***' 0.001 '**' 0.01 '*' 0.05 '.' 0.1 ' ' 1				
Residual standard error: 637.5 on 15 degrees of freedom				
Multiple R-squared: 0.5412				
Adjusted R-squared: 0.3577				
F-statistic: 2.949 on 6 and 15 DF **p-value: 0.04172**				

Table 3. Evolution of the Correlation between LEED certified buildings presence and the increase of the real estate market value

Year	Correlation value
2015	0.6098
2017	0.6488
2019	0.6945
2021	0.6915
2023	0.6959

f_2 = White population (non-Hispanic white) (%);
f_3 = Population with bachelor's degree (%);
f_4 = Residential market value;
f_5 = Renters (%);
f_6 = Residents below poverty line (%);
f_7 = Household income (%);

The statically significance threshold established for the model is with p-value at least less than 0.1 (typically ≤ 0.1) which provides at least moderate evidence against the null hypothesis. To obtain the importance of each variable the authors use the Random Forest method supported again with R software. to obtain a closer approach to the real weights of variables to the final price.

The first regression which includes the seven main variables associated with gentrification (see Table 4), shows an effective relationship between the variables and the concentration of LEED-certified buildings. This is demonstrated by the R squared equal to 0.56 for which 56% of the phenomena is explained by these variables suggesting that the model is a moderate fit to the data. Above all, the p-value is equal to 0.048 which indicates a strong level of significance. Therefore, the null hypothesis according to which these variables affect the level of concentration of LEED-certified buildings can be accepted, since the p-value is above the significance level of ≤ 0.1 (Table 4).

Table 4. Summary of the factors that define gentrification

Coefficients:				
	Estimate	Std. Error	t value	Pr(>|t|)
(Intercept)	−6.153e + 00	4.799e + 01	−0.128	0.8997
F1	−1.252e + 00	2.626e + 00	−0.477	0.6402
F2	3.426e-01	7.178e-01	0.477	0.6400
F4	7.463e-05	2.952e-05	2.528	0.0232
F6	3.504e-01	1.451e + 00	0.242	0.8124
F3	−1.459e + 00	1.476e + 00	−0.988	0.3386
F7	−4.400e-06	3.080e-04	−0.014	0.9888
F5	−1.756e-02	6.467e-01	−0.027	0.9787
Signif. Codes: 0 '***' 0.001 '**' 0.01 '*' 0.05 '.' 0.1 ' ' 1				
Residual standard error: 24.42 on 15 degrees of freedom				
Multiple R-squared: 0.56				
Adjusted R-squared: 0.3547				
F-statistic: 2.728 on 7 and 15 DF **p-value: 0.04873**				

Nevertheless, the authors want to depurate more the model and find the more influential variables. The depuration of the model identifies the five most relevant variables (see Table 5). Consequently, the model was adapted as follows.

$$Y_b = \beta_0 + \beta_1 f_1 + \beta_2 f_2 + \beta_3 f_3 + \beta_4 f_4 + \beta_6 f_6 + u_i$$

The selected variables are:

f_1 = Distance to downtown (km);
f_2 = White population (non-Hispanic white) (%);
f_3 = Population with bachelor's degree (%);

f_4 = Residential market value;
f_6 = Residents below poverty line (%);

Table 5. Summary of the factors that define gentrification

Coefficients:				
	Estimate	Std. Error	t value	Pr(>|t|)
(Intercept)	−6.989e + 00	4.166e + 01	−0.168	0.8688
F1	−1.296e + 00	1.896e + 00	−0.684	0.5034
F2	3.194e-01	5.059e-01	0.631	0.5362
F4	7.468e-05	2.771e-05	2.695	0.0153 *
F6	3.507e-01	1.307e + 00	0.268	0.7917
F3	−1.419e + 00	1.136e + 00	−1.249	0.2286
Signif. Codes: 0 '***' 0.001 '**' 0.01 '*' 0.05 '.' 0.1 ' ' 1				
Residual standard error: 22.94 on 17 degrees of freedom				
Multiple R-squared: 0.56				
Adjusted R-squared: 0.4305				
F-statistic: 4.327 on 5 and 17 DF **p-value: 0.01009**				

The second regression, which includes the five main variables, shows an effective relationship between the variables and the price of the buildings. This is demonstrated by the R squared that remains equal to the previous model with 0.56, for which 56% of the phenomena is explained by these variables, suggesting that there is a moderate fit to the data in the model. Above all, the p-value is equal to 0.001, which indicates a very strong level of significance. Therefore, the null hypothesis according to which these variables affect the level of LEED-certified buildings distribution can be accepted since the p-value is above the significance level of ≤ 0.1.

Using the Random Forest package, it was possible to obtain the importance of each variable for the second model. The results show that the most relevant variable is F5, which is the physical distance to downtown, represents 33.35%, followed by F4, which is the average residential market value that represents 30.40%; next is F3, which is the population with a bachelor's degree with the 13.98%. This means that three variables explain 77.73% of the model. The other two variables with less influence in the model are F2, which is the White population (non-Hispanic white) with 11.78%, and finally, F6, which is Residents below the poverty line and has a lesser value with 10.46%. The second model shows a more equilibrated influence of the variables than the first one. According to the authors, these five variables can explain green gentrification in the city of Boston (see Table 6).

Table 6. Green Gentrification factors values for each Boston neighbourhood (2021)

Neighbourhood	LEED buildings	Distance (km)	White population (%)	Market value (USD)	Poverty (%)	Education (%)
Allston	25	5.85	54.1	608406	4.1	42.3
Back Bay	37	1.99	70.7	1173621	1.9	43.4
Beacon Hill	2	0.89	86	911907	0.6	41.9
Brighton	7	8.06	62.4	551935	0.5	40.6
Charlestown	5	2.68	70.5	848139	2	35
Chinatown	9	0.81	22.3	952843	0.9	25.6
Dorchester	7	6.36	22.5	595495	21.7	18.5
Downtown	142	0.00	63.7	1416687	1.9	26.2
East Boston	27	3.94	34.5	585023	6.2	21.6
Fenway	23	3.39	55.6	624483	6.2	34.1
Hyde Park	2	12.52	22.3	532586	4.3	21.1
Jamaica Plain	7	7.11	53.5	697053	5.4	27.6
Longwood	13	4.34	68.6	1120721	0.1	40
Mattapan	1	9.16	5.2	490032	3.9	13
Mission Hill	2	4.55	41.2	766925	4.9	28.3
North End	1	1.14	85.3	800465	0.8	50.6
Roslindale	1	9.91	48.7	617245	2.8	24.3
Roxbury	16	4.76	14.8	560826	12.5	17.7
South Boston	23	2.53	77	816170	3.8	44.4
South Boston Waterfront	59	1.66	81.9	1374864	0.1	39.6
South End	33	1.94	53.8	999212	5.7	28.2
West End	13	1.20	65.3	591490	0.8	36.8
West Roxbury	1	11.85	65.1	663589	1.9	29

4.3 Green Gentrification Index for the City of Boston

Once the variables and the weights of each of them are established, it is possible to compose the index to approximate the risk of green gentrification of each of the 23 neighbourhoods of the city of Boston (see Table 7). The values of each variable were standardised on a scale of 1 to 5, where 1 is the highest risk of green gentrification, and 5 is the lowest risk and is subsequently multiplied by the weight of importance obtained using the Random Forest methodology. Finally, it is possible to obtain the comprehensive score of the Green Gentrification Index for the city of Boston at the neighbourhood level.

Table 7. Green Gentrification Index results per neighbourhood (2021)

Neighborhood	LEED buildings	Distance 33.35%	White population 11.78%	Market value 30.40%	Poverty 10.46%	Education 13.98%	Total
Allston	25	3	3	2	5	5	3
Back Bay	37	5	4	5	5	5	5
Beacon Hill	2	5	5	4	5	5	5
Brighton	7	2	4	1	5	5	3
Charlestown	5	4	4	4	5	4	4
Chinatown	9	5	2	4	5	3	4
Dorchester	7	3	2	2	1	2	2
Downtown	142	5	4	5	5	3	5
East Boston	27	4	2	2	4	3	3
Fenway	23	4	3	3	4	4	4
Hyde Park	2	1	2	1	5	3	2
Jamaica Plain	7	3	3	3	4	3	3
Longwood	13	4	4	5	5	5	5
Mattapan	1	2	1	1	5	2	2
Mission Hill	2	4	3	3	5	3	4
North End	1	5	5	4	5	5	5
Roslindale	1	2	3	2	5	3	3
Roxbury	16	4	1	2	3	2	3
South Boston	23	4	4	4	5	5	4
South Boston Waterfront	59	5	5	5	5	4	5
South End	33	5	3	5	4	3	4
West End	13	5	4	2	5	4	4
West Roxbury	1	1	4	3	5	3	3

The results of the index (see Fig. 7) show that the most vulnerable neighbourhoods to green gentrification and require special attention from the public authorities to invest in public green buildings and other types of green infrastructures. These neighbourhoods are Dorchester, Mattapan and Hyde Park, which are traditionally inhabited by historically vulnerable groups and are more distant from the city centre, where economic activity and green buildings are concentrated. Also, traditional peripheral neighbourhoods such as Roxbury and Jamaica Plain have a moderate risk that should raise awareness to public authorities and civil society to reclaim more public investments for green practices, which include green-certified buildings. The least vulnerable neighbourhoods are the ones that surround the downtown area and financial centre. In these neighbourhoods, the authors believe that the private sector is taking the initiative to transform the existing building stock into a greener one because of the economic benefits it provides.

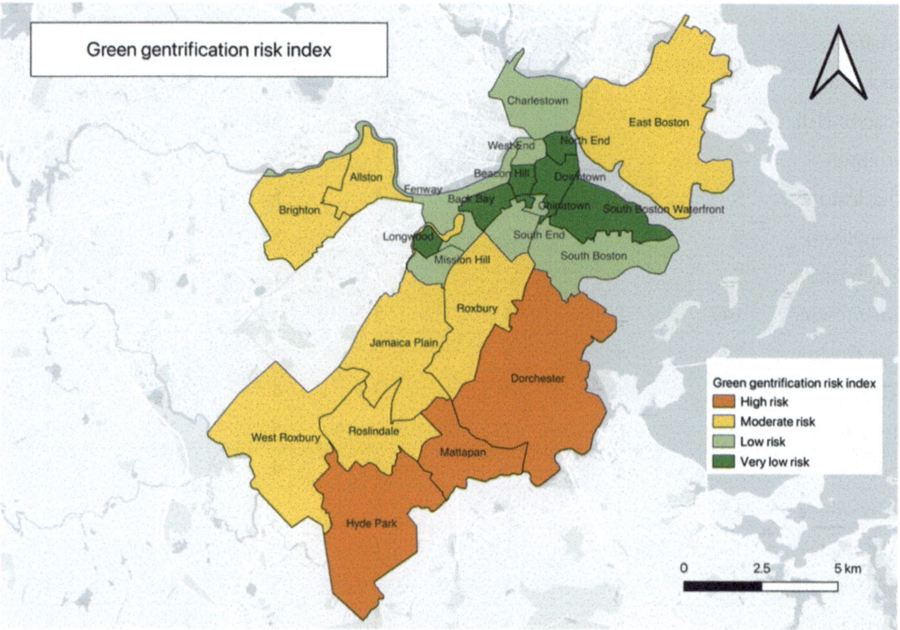

Fig. 7. Green gentrification index at the neighbourhood level for the city of Boston (2021)

5 Discussion and Conclusions

Urban planning plays a pivotal role in the adoption of green buildings in cities. The integration of sustainable and high-performance building practices into urban planning can contribute to the overall environmental, social, and economic goals of a city. The inequalities in the distribution of green-certified buildings in market-driven contexts are one of the main challenges that cities can face in the transformation of the conventional building stock into a greener one. In Boston, most green-certified buildings are concentrated in the wealthiest and centric neighbourhoods of the city, where people can afford to pay the cost premium that implies the development of a green-certified building. This situation leaves the inhabitants of low-income and traditional vulnerable neighbourhoods behind in the transformation of their building stock and in the overall urban transition to greener and more sustainable cities. To cope with this challenge, it is recommended to stimulate the presence of green-certified public buildings and housing in vulnerable neighbourhoods. This can help to reduce the risk of green gentrification and achieve a more egalitarian urban transition. The paper intends to contribute to a better understanding of this challenge and propose a set of recommendations based on their findings.

Firstly, there is a clear synergy between green-certified building criteria such as LEED with urban transition values and goals. This makes the implementation of green-certified buildings a key player in urban transformation initiatives such as the Global Covenant of Mayors. There is a clear influence of the LEED-certified evaluation categories which are closely related to urban transition values in the final price of buildings in the city of Boston. Consequently, the price difference between conventional and green buildings can be considered as the costs of the urban transition, in the first instance in the construction sector and eventually in general. The results of the stepwise regression helped to identify the five main factors that determine the distribution of LEED Certified buildings in Boston, which are Distance to downtown (km), White population (non-Hispanic white) (%), Population with bachelor's degree (%), Residential market value, and Residents below poverty line (%). The identified factors were associated with green gentrification and used to create a green gentrification risk index for the city of Boston. This means that the neighbourhoods with the lowest values are the ones more vulnerable to green gentrification, and their population has a higher risk of being displaced at the moment they cannot afford the costs of living in areas where the building stock has been replaced by green certified buildings developed by private investors. Using the stepwise regression model, it was possible to identify the influence of each variable on the model to subsequently set the weights for each variable to be used in the index. The first approach of the index spatially identifies the most vulnerable neighbourhoods to green gentrification and can be useful for decision-makers to focus their efforts on investing in public green buildings and housing or facilitate their presence with tax and regulatory stimulus. In the case of Boston, the most vulnerable neighbourhoods to green gentrification are located in the peripheries of the city, mainly populated by historically vulnerable groups and with a very low presence of green-certified buildings. The article expects to collaborate to consider the risks of green gentrification in their urban transformation process in other cities of the United States or even in other countries. Also, it calls for a bigger involvement of the public sector in the renovation of the building stock and not leave the transformation to green buildings just to the private sector.

Acknowledgments. This research work is the result of the research activities conducted within the following PNRR Research Project funded by the European Union – NextGenerationEU: "ECO-SET - A Multidisciplinary approach to plan Ecosystem Services for cities in Transition" research project, Finanziato dall'Unione europea- Next Generation EU, Missione 4 Componente 2 CUP B53D23029450001.

References

Boston Housing Authority 2023: Public Housing Community Map (2023). https://Bostonhousing.Org/En/Housing-Communities/Public-Housing-Communities/Public-Housing-Map.Aspx

Chen, S.-W., Gou, Z.: Spatiotemporal distribution of green-certified buildings and the influencing factors: a study of U.S. Heliyon **9**(11), e21868 (2023). https://doi.org/10.1016/j.heliyon.2023.e21868

Global Covenant of Mayors. «Boston, MA Global Covenant of Mayors». Accessed September 14 2023 (2023). https://www.globalcovenantofmayors.org/cities/usa/united-states/boston-ma/

Fastenrath, S., Braun, B.: Sustainability transition pathways in the building sector: energy-efficient building in Freiburg (Germany). Appl. Geogr. **90**(January), 339–349 (2018). https://doi.org/10.1016/j.apgeog.2016.09.004
Wen, B., et al.: The role and contribution of green buildings on sustainable development goals. Build. Environ. **185**(November), 107091 (2020). https://doi.org/10.1016/j.buildenv.2020.107091
Affolderbach, J., Schulz, C.: Green building as urban Sustainability Transitions. In The Urban Book series, pp. 15–27 (2018). https://doi.org/10.1007/978-3-319-77709-2_2
Sokolowski, I.L., Maheshwari, A., Malik, A.: Green Buildings: A Finance and Policy Blueprint for Emerging Markets. International Finance Corporation (2019). http://documents.worldbank.org/curated/en/586841576523330833/Green-Buildings-A-Finance-and-Policy-Blueprint-for-Emerging-Markets
Tsai, I.-C.: Value capitalization Effects of green buildings: a new insight through time trends and differences in various price levels. Building and Environment **224**(October), 109577 (2022). https://doi.org/10.1016/j.buildenv.2022.109577
BreGroup. All about BREEAM - BRE Group». BRE Group - Building a Better World Together. January 5 2023. Accessed 18 Oct 2023 (2023). https://bregroup.com/buzz/all-about-breeam/#:~:text=BREEAM%20is%20an%20internationally%20recognised,more%20than%2074%20other%20countries
U.S. Green Building Council. LEED Hits 100,000 Certified Projects. 2022. U.S. Green Building Council. October 10 2022. Accessed 18 Oct 2023 (2022). https://www.usgbc.org/articles/leed-hits-100000-certified-projects
Goubran, S., Walker, T., Cucuzzella, C., Schwartz, T.: Green building standards and the United Nations' sustainable development goals. J. Environ. Manage. **326**(January), 116552 (2023). https://doi.org/10.1016/j.jenvman.2022.116552
European Union. Driving Urban Transitions to a Sustainable Future Roadmap. 5908. Driving Urban Transition Partnership. Driving Urban Transition Partnership. Accessed 15 Apr 2023 (2022). https://dutpartnership.eu/wp-content/uploads/2022/09/DUT-Roadmap-2022-komprimiert.pdf
Haldrup, K., Snällfot, D.: On financing of urban transition viewed from the Oresund area: when the political agenda of urban transition meets the market. Sustainability **6**(12), 9057–9079 (2014). https://doi.org/10.3390/su6129057
Dwaikat, L.N., Ali, K.N.: Green buildings cost premium: a review of empirical evidence. Energy and Buildings **110**(January), 396–403 (2016). https://doi.org/10.1016/j.enbuild.2015.11.021
Cushman, W.: Green is good: sustainable office outperforms in class a urban markets. Cushman & Wakefield Research Spotlight. Accessed 9 Nov 2023 (2021). https://cw-gbl-gws-prod.azureedge.net/-/media/cw/americas/united-states/insights-pdfs/green-is-good-spotlight_final1.pdf?rev=e26315797d7d49faa6c58ca7762f91a6
Sax, D., Nesbitt, L., Quinton, J.: Improvement, not displacement: a framework for urban green gentrification research and practice. Environmental Science & Policy **137** (november), 373–83 (2022). https://doi.org/10.1016/j.envsci.2022.09.013
Gupta, R.: Green Building Interventions for Social Housing. UN-Habitat, Nairobi (2015)
European Commission. Going Climate-Neutral by 2050 – A Strategic Long-Term Vision for a Prosperous, Modern, Competitive and Climate-Neutral EU Economy. Publications Office (2019). https://doi.org/10.2834/02074
Peng, Y., Bai, X.: Financing urban low-carbon transition: the catalytic role of a city-level special fund in Shanghai. J. Cleaner Production **282**(february), 124514 (2021). https://doi.org/10.1016/j.jclepro.2020.124514
Liu, T., et al.: Sustainability considerations of green buildings: a detailed overview on current advancements and future considerations. Sustainability **14**(21), 14393 (2022). https://doi.org/10.3390/su142114393

Darko, A., Chan, A.P.C., Ameyaw, E.E., He, B.-J., Olanipekun, A.O.: Examining issues influencing green building technologies adoption: the United States green building experts' perspectives. Energy and Buildings **144**(June), 320–332 (2017). https://doi.org/10.1016/j.enbuild.2017.03.060

Simpeh, E.K., Smallwood, J., Ahadzie, D.K., Mensah, H.: Analytical taxonomy of challenges to the implementation of green building projects in South Africa. Int. J. Constr. Manag. **23**(2), 286–296 (2021). https://doi.org/10.1080/15623599.2020.1863172

Uğur, L.O., Leblebici, N.: An examination of the LEED Green Building certification system in terms of construction costs. Renew. Sustain. Energy Rev. **81**(January), 1476–1483 (2018). https://doi.org/10.1016/j.rser.2017.05.210

Chakravarthy, P., Suganya, R., Nivedhitha, M., Parthiban, A., Sivaganesan, S.: Barriers and project management practices in green buildings. Materials Today: Proceedings **52**(January), 1131–1134 (2022). https://doi.org/10.1016/j.matpr.2021.11.007

Ayarkwa, J., Opoku, D.-G., Antwi-Afari, P., Li, R.Y.M.: Sustainable building processes' challenges and strategies: the relative important index approach. Cleaner Eng. Technol. **7**(April), 100455 (2022). https://doi.org/10.1016/j.clet.2022.100455

Çiner, F., Doğan-Sağlamtimur, N.: Environmental and sustainable aspects of green building: a review. IOP Conference Series: Materials Science and Eng. **706**(1), 012001 (2019). https://doi.org/10.1088/1757-899x/706/1/012001

Zuniga-Teran, A.A., Gerlak, A.K., Elder, A.D., Tam, A.: The unjust distribution of urban green infrastructure is just the tip of the iceberg: a systematic review of place-based studies. Environ. Sci. Policy **126**, 234–245 (2021). https://doi.org/10.1016/j.envsci.2021.10.001

Mady, A.A.: Green building rating systems as a design target for public buildings: a case study analysis. Prog. Ind. Ecol. Int. J. **11**(2), 118 (2017). https://doi.org/10.1504/PIE.2017.088847

JPI Urban Europe. Preparation of the European Partnership Driving Urban Transitions Report on the AGORA Strategic Dialogues. Urban Europe. September 2020. Accessed 12 Apr 2023 (2020). https://jpi-urbaneurope.eu/wp-content/uploads/2020/10/DUT-AGORA-online-1.pdf

Floater, G., Dowling, D., Chan, D., Ulterino, M., Braunstein, J., McMinn, T.: Financing the Urban Transition: Policymakers' Summary. Coalition for Urban Transitions. London and Washington, DC (2017). http://newclimateeconomy.net/content/cities-working-papers

Tatari, Ö., Küçükvar, M.: Cost premium prediction of certified green buildings: a neural network approach. Build. Environ. **46**(5), 1081–1086 (2011). https://doi.org/10.1016/j.buildenv.2010.11.009

Kim, J.-L., Greene, M., Kim, S.; Cost comparative analysis of a new green building code for residential project development. J. Construction Division and Management **140**(5) (2014). https://doi.org/10.1061/(asce)co.1943-7862.0000833

Cole, H., Lamarca, M.G., Connolly, J.J., Anguelovski, I.: Are green cities healthy and equitable? Unpacking the relationship between health, green space and gentrification. J. Epidemiology and Community Health, August, jech-209201 (2017). https://doi.org/10.1136/jech-2017-209201

Rigolon, A., Németh, J.: Green gentrification or 'just green enough': Do park location, size and function affect whether a place gentrifies or not? Urban Studies **57**(2), 402–20 (2019). https://doi.org/10.1177/0042098019849380

Taylor, D.E.: Introduction: the evolution of environmental justice activism, research, and scholarship. Environ. Pract. **13**(4), 280–301 (2011). https://doi.org/10.1017/S1466046611000329

Ferguson, M., Roberts, H., McEachan, R., Dallimer, M.: Contrasting distributions of urban green infrastructure across social and ethno-racial groups. Landsc. Urban Plan. **175**(July), 136–148 (2018). https://doi.org/10.1016/j.landurbplan.2018.03.020

Shonkoff, S.B.C., Morello-Frosch, R., Pastor, M., Sadd, J.: The climate gap: environmental health and equity implications of climate change and mitigation policies in California—a review of the literature. Clim. Change **109**(S1), 485–503 (2011). https://doi.org/10.1007/s10584-011-0310-7

U.S. Green Building Council. What is LEED certification? Accessed 6 May 2023 (2022). https://support.usgbc.org/hc/en-us/articles/4404406912403-What-is-LEED-certification-

Pushkar, S.: LEED 2009 recertification of existing buildings: bonus effect. Sustainability **13**(19), 10796 (2021). https://doi.org/10.3390/su131910796

Amiri, A., Ottelin, J., Sorvari, J.: Are LEED-certified buildings energy-efficient in practice? Sustainability **11**(6), 1672 (2019). https://doi.org/10.3390/su11061672

U.S. Green Building Council. LEED 2009 for Existing Buildings. USGBC, Washington D.C (2010)

Priya, A.: Case study methodology of qualitative research: key attributes and navigating the conundrums in its application. Sociological Bulletin **70**(1), 94–110 (2020). https://doi.org/10.1177/0038022920970318

City of Boston. Welcome - Analyze Boston (2023). Accessed 12 Sep 2023. https://data.boston.gov/

Boston Planning and Development Agency «Research Publications | Boston Planning & Development Agency». 2023 BPDA. Accessed 12 Sep 2023. http://www.bostonplans.org/research/research-publications

United States Census Bureau QuickFacts. U.S. Census Bureau QuickFacts: Boston City, Massachusetts». Census Bureau QuickFacts (2022). Accessed 21 May 2023. https://www.census.gov/quickfacts/fact/table/bostoncitymassachusetts/PST045223

U.S. Green Building Council. USGBC | U.S. Green Building Council (2023). Accessed 31 Oct 2023. https://www.usgbc.org/

The Green Building Information Gateway. Boston, Massachusetts: GBIG. GBIG (2023). Accessed 20 Apr 2023. https://www.gbig.org/places/3135

Zillow. Boston, Massachusetts. Zillow. Accessed: 20 Apr 2023 (2023). https://www.zillow.com/boston-ma/

Commercial Café. Boston, Massachusetts. Commercial café (2023). Accessed 20 Apr 2023. https://www.commercialcafe.com/commercial-real-estate/us/ma/boston/

City of Boston. Climate Action Plan. City of Boston 2019 Climate Action Plan (2019). Accessed 23 Apr 2023. https://www.boston.gov/sites/default/files/embed/file/2019-10/city_of_boston_2019_climate_action_plan_update_4.pdf

City of Boston. Imagine Boston. City of Boston (2018). Accessed 15 Oct 2023. https://www.boston.gov/sites/default/files/embed/file/2018-06/imagine20boston202030_pages2.pdf

Global Covenant of Mayors. Boston is Taking Action on Climate Change; Global Covenant of Mayors (2023). Accessed 13 Apr 2023. https://www.globalcovenantofmayors.org/cities/usa/united-states/boston-ma/

CDP. Cities Scores – CDP (2022). Accessed 15 Apr 2023. https://www.cdp.net/en/cities/cities-scores

Metropolitan Area Planning. Council«DataCommon» (2023). Accessed 21 May 2023. https://datacommon.mapc.org/browser/datasets/316

Boston Planning & Development Agency Research Division. (2017). Neighborhood Profiles. Accessed 20 May 2023. http://www.bostonplans.org/getattachment/7987d9b4-193b-4749-8594-e41f1ae27719

Boston Planning & Development Agency Research Division (2019). Neighborhood Profiles. Accessed 20 May 2023. http://www.bostonplans.org/getattachment/7987d9b4-193b-4749-8594-e41f1ae27719

Boston Planning & Development Agency Research Division (2021). Neighborhood Profiles. Accessed 20 May 2023. http://www.bostonplans.org/getattachment/7987d9b4-193b-4749-8594-e41f1ae27719

Boston Planning & Development Agency Research Division (2023). Neighborhood Profiles. Accessed 20 May 2023 http://www.bostonplans.org/getattachment/7987d9b4-193b-4749-8594-e41f1ae27719

National Neighborhoods Indicators Partnership. Guide to Measuring Neighborhood Change TO Understand and Prevent Displacement. NNIP, United States (2017)

National Community Reinvestment Coalition. Gentrification and Disinvestment 2020. Do Opportunity Zones Benefit or Gentrify Low-Income Neighborhoods? NCRC, United States (2020)

Vlaams Energiebedrijf: About This Good Practice (2019). Accessed 21 May 2023. https://www.interregeurope.eu/good-practices/vlaams-energiebedrijf-veb-flemish-energy-company#further-Information

Measuring the Value of Ecosystem Services

Raffaella Campanella(✉)

Mediterranean University of Reggio Calabria, Reggio Calabria, Italy
rcampanella@unirc.it

Abstract. Although the literature on the valuation of ecosystem services is becoming substantial, with an increase in the number of scientific articles over the last thirty years, it still lacks a widely accepted methodology for mapping and measuring the value of ecosystem services that is suitable for defining a strategy for their spatial planning and management. Measuring the value of ecosystem services requires, in fact, the coordination of multiple, composite and, quite often, conflicting dimensions of value. It is an articulated process, as the valuation of each ecosystem service requires scientific expertise in different fields that must, however, interact with each other, as well as a wealth of data whose availability and accessibility varies in terms of accuracy, updating, size scale, etc. Moreover, while for the assessment of some ecosystem services, well-established methodologies are available, for others, assessment techniques are still being tested. This paper - which summarises part of the work carried out during the first year of the Pilot Project "Climate adaptation plans for the reduction of the ecological footprint and ecological debt aimed at improving the conservation" within the SPOKE 4 (executor) "Resilience and accessibility in the valorisation of local cultural and natural heritage" (for a Budget of PP 2.2 million Euro) of the TECH4YOU Research Project "Technologies for climate change adaptation and quality of life improvement". PNRR-funded project - explores the main theories and experiences in this field.

Keywords: Ecosystem Services · Assessment Methods · Indicators and Mapping

1 Foreword

This paper summarises some of the work carried out during the first year of Goal 4.6 "Climate change planning to promote cultural and natural heritage: Demand-driven ecosystem services based on ICT and AI enabling technologies", and of the Pilot Project "Climate adaptation plans for the reduction of the ecological footprint and ecological debt aimed at improving the conservation" within SPOKE 4 (executor) "Resilience and accessibility in the enhancement of local cultural and natural heritage" (for a PP Budget 2.2 million Euro) of the TECH4YOU Research Project "Technologies for climate change adaptation and quality of life improvement". Project financed PNRR mission 4, component 2 investment 1.5 - Innovation Systems 2 - ranked 1st, at national level, for proposal quality.

F. Moraci et al. (Eds.): NMP 2024, LNNS 1272, pp. 62–79, 2025.
https://doi.org/10.1007/978-3-031-82927-7_4

In particular, it refers to a specific segment of Action 1 - Green and blue infrastructures as mitigation and offset ecosystem services, namely that aimed at defining the measurement of the value of ecosystem services.

1.1 Ecosystem Services

Ecosystem services can be defined, in a very general way, as the services provided by nature that are necessary for the sustenance of human beings and the societies of which they are part. The concept of ecosystem services originated in the scientific literature of the 1970s and spread from the 1990s onwards with the aim of giving economic weight to the contribution that ecological processes make to human well-being. To the traditional strand of the study of ecosystem services concerning the territory and related to natural ecosystems has recently been added that of urban ecosystem services, which is related to Urban Ecology, the latter generally defined as a sub-discipline of ecology. This approach, which has emerged fairly recently, has its roots in studies conducted since the 1920s in the United States and after World War II in Europe, but emerges as a discipline mainly at the end of the 20th century, with contributions such as Bolund & Hunhammar's article, 'Ecosystem services in urban areas'. (1999).

Over the last two decades, ecosystem services have therefore been the subject of in-depth conceptual and methodological analysis and evaluation models that have been continually refined, including the territorial representation of these services and the quantification of their value, both monetary and otherwise, up to the introduction of Payment for Ecosystem Services (PES) mechanisms.

This report provides, without claiming to be exhaustive, an overview of the classification and approaches aimed at their analysis and evaluation, including their cartographic restitution.

In particular, we consider the dimension of territorialisation of ecosystem services with associated qualitative or even quantitative attribution, in the second case not necessarily economic, to be used above all, but not only, as a tool for defining effective mitigation and compensation measures and launching virtuous policies for the construction of green and blue infrastructures and urban and territorial regeneration.

With regard to the valuation of ecosystem services, which will be dealt with in more detail below, it is noted that the recognised methods are both monetary (these include market methods, for example of greenhouse gases, but also exchange methods, such as PES) and non-monetary, in the second case with the objective of representing quantity and quality not expressed in money but in such a way as to make the different values evident; among the latter methods are also socio-cultural ones.

The concept of ecosystem services took off, in parallel with Urban Ecology (a discipline for the study of urban ecosystems), in the 1980s, and the first publication containing this term, according to the majority of studies on the subject, was entitled 'Extinction: the Causes and Consequences of the Disappearance of Species' (1981), by the American biologists Paul R. Ehrlich and Anne H. Ehrlich.

With the studies by Robert Costanza (1997) and Gretchen C. Daily (1997) began the dissemination of this concept, which was consolidated with the essays of Rudolf S. De Groot (2002), the Millennium Ecosystem Assessment (MEA 2005) and the TEEB Report (2010).

The scientific literature and the documents produced provide different definitions and classifications of ecosystem services, although they are anchored to a rather clear conceptualisation and linked to an approach that puts the human being at the centre; in other words, the dynamics of ecosystems are read in function of the satisfaction of human needs. Several conceptual schemes are proposed through which to represent, in a synthetic and structured manner, ecosystem services and to relate ecosystem functions to goods and ecosystem services themselves, and thus, ecosystems to socio-economic systems.

2 Ecosystem Service Assessment Methods

Although the literature concerning the valuation of ecosystem services is becoming substantial, with an increase in the number of scientific articles since 1997 - the year of publication of the work of Robert Costanza who was among the first to bring this issue to the forefront - there is still a lack of a widely accepted methodology aimed at mapping and measuring the value of ecosystem services that is suitable for defining a strategy for their spatial planning and management.

Measuring the value of ecosystem services requires, in fact, the coordination of multiple, composite and, quite often, conflicting dimensions of value. It is an articulated process, as the valuation of each ecosystem service requires scientific expertise in different fields that must, however, interact with each other, as well as a wealth of data whose availability and accessibility varies in terms of accuracy, updating, size scale, etc.

Moreover, while for the assessment of some ecosystem services, fairly well-established methodologies are available, for others, assessment techniques are still being tested.

In Italy, for example, the studies conducted are rather limited, and almost always circumscribed in terms of the extent of the area considered and the number of ecosystem services examined. Measuring the value of ecosystem services, in fact, is a fairly recent line of research (Santolini R., 2008). And, specifically, it seems to lack an exhaustive and in-depth definition of the various ecosystem services at the most detailed scales, such as regional and urban, which would instead be of great use to be able to direct planning and management strategies towards more conscious choices for a sustainable use of resources. (Santolini R., Morri E., 2017a; Santolini R., Morri E., 2017b). On the other hand, it is correct to point out that the assessment of ecosystem services is not a simple and quite costly operation, particularly due to the unavailability of suitable and up-to-date databases at the various scales of intervention.

To date, the mapping and valuation of ecosystem service provision hinges mainly on literature studies and modelling methods. Among these, the main ones concern: the market valuation method (Bateman et al. 2002; de Groot et al. 2010), ecological process simulation (Nedkov, 2012; Stürck et al. 2014), InVEST models (Arcidiacono et al. 2015; Boithias et al. 2014). Also of great interest for our research is the methodological approach based on land cover (Burkhard et al. 2015; Burkhard et al. 2009; Burkhard et al. 2012). The latter method, which is widely used, consists of quantitatively assessing the ecosystem service provisioning capacity of different land cover types.

2.1 Assessing Ecosystem Services Through the Economic Approach

The public good character of most ecosystem services means that, very often, they do not have an explicit economic value, which creates significant difficulties in the valuation process.

Robert Costanza, for one, points out that the methodical underestimation of the environmental issue in decision-making processes can be traced back to the fact that the goods and services provided by natural capital cannot be estimated in comparable terms with other services and other forms of capital.

It is, however, now well established that the loss of ecosystem services is a threat not only to our existence but also to our economy. Therefore, the issue is the subject of various studies and research.

In this regard, the following are some highlights from a study conducted by the Institute of Management of the Scuola Superiore Sant'Anna in Pisa on the economic valuation of ecosystem services.

Why Evaluate Ecosystem Services

- Biodiversity suffers from the effects caused by indirect factors, such as population growth, a lack of knowledge of the issues surrounding it and the fact that its economic value is not considered in decision-making processes.
- The EU biodiversity target is based on the recognition that, in addition to its intrinsic value, biodiversity and the services it provides have considerable economic value that the market rarely captures

Preserving and restoring ecosystems and their services implies that the economic value of Ecosystem Services is recognised and its valuation integrated into decision-making processes at both public and private levels.

Economic valuation helps to strengthen understanding and communication of the values of natural capital and ecosystem services.

What to evaluate

- Direct use value. Use value derives from the actual, expected or possible consumption of a good. When there is direct human use, we speak of direct use value: it is used for all provisioning services and some cultural services, e.g. recreation.
- Indirect use value. Usage can also occur indirectly, and this is the case with the benefits derived from all regulation services
- Option valuc. Option value falls into the category of use value, but in this case the use is not current but future. The individual does not use the good/service currently but is nevertheless willing to pay a certain amount in order to have the possibility of using that good/service in the future. The definition of option value is linked to the concept of risk aversion: since people are uncertain about the future supply of the environmental good/service, they are willing to pay a sum of money today that guarantees them the possibility of use in the future.
- Existence value. It is based on the awareness of individuals that the considered good exists and continues to exist, even if they will never use it. An individual may be willing to pay for the preservation of the Amazon rainforest even though they will never visit it, but only because they value its existence.

- Inheritance value. It derives from the fact that the person valuing is willing to pay a certain amount so that other people may enjoy the property in question.

How to evaluate

There are two different approaches to which different methodologies can be attributed:

1. Direct or 'market price' methods are based on measuring the value of SEs through market prices. The market, in fact, is the place where individuals reveal their preferences and willingness to pay (one speaks in this case of '*revealed preference*').
2. Indirect or 'contingent valuation' methods involve field surveys to identify the willingness to pay for a given ecosystem service (hence '*stated preference*'). Many ecosystem services, however, are not traded in observable markets, and are not closely related to any marketed products. Thus, individuals cannot 'reveal' how much they are willing to pay through their market purchases or through their actions.

Some methods are more suitable than others for defining the value of particular ecosystem services. For example, market prices are often used to estimate provisioning services; while surveyed preferences are usually used to define non-use values. In some contexts, however, it may be convenient to use several methodologies simultaneously.

Non-market methods are more widely used, as they allow the impact of a hypothetical physical transformation of the ecosystem on the well-being of individuals to be measured; in this way they provide an ex-ante estimate, which is useful for defining possible intervention scenarios. Such methods, however, are reproached for the predominance of users' opinions and needs over the maintenance or improvement of ecological conditions.

These two macro-categories are not, however, the only ones that can be used; in fact, there are others such as meta-analysis and value transfer, which, however, cannot be considered true economic evaluation tools, as they use data from other studies.

2.2 Assessing Ecosystem Services Through the Land Use Approach

The aforementioned approach, based on the quantitative assessment of the provisioning capacity of ecosystem services by different land cover types, is a widely used assessment methodology. In recent years, many researchers have used this method for both large- and small-scale assessments.

In this regard, the work of Burkhard et al. has been widely recognised for its advantages over other methods. It is, in fact, a fairly rapid evaluation method with clear advantages for decision-making. Furthermore, it can be applied to different spatial situations and requires little data at an early stage: land cover types and expert knowledge.

Indeed, the quality and quantity of ecosystem services are explicitly conditioned by structural and functional changes in different land cover types (ecosystems). For example, changes in land use can affect regional or global climate (Pielke, 2002) which, in turn, can lead to a decrease in the capacity of local and global climate regulating services. Additionally, improper land use can cause habitat and landscape fragmentation (Mitchell et al. 2015; Mitchell et al. 2013), the subsequent disintegration of wildlife habitats as well as the modification of landscape connectivity. And all this can affect, directly or indirectly, the formation and provision of ecosystem services.

However, it must also be said that in some cases, the impact of land use change on ecosystem services can also be positive. For example, Lovell (2010) argues that urban agriculture is a sustainable and multifunctional land-use option for cities; Hostetler et al. (2011) assert that the creation of urban green infrastructure is crucial for the maintenance of biodiversity. Therefore, proper management and optimisation of land use patterns can ensure the maintenance and increase the genesis of new ecosystem services (de Groot et al. 2010).

2.3 Assessing the Socio-cultural Value of Ecosystem Services

People attribute different material, moral, spiritual, aesthetic and other values to the environment; their values can influence their attitudes and actions toward ecosystems and the services they provide. These values include emotional, affective and symbolic visions related to urban nature that in most cases cannot be adequately captured by commodity translates and monetary metrics. Here we refer to these values broadly as social and cultural values. The literature on ecosystem services has defined cultural values as "aesthetic, artistic values, educational, spiritual and/or scientific benefits of ecosystems" (Costanza et al. 1997) or as "non- material benefits that people gain from ecosystems through spiritual enrichment, cognitive development, reflection, recreation and aesthetic experience" (Millennium Ecosystem Assessment, 2005).

Social and cultural values are included in all major types of ecosystem services (Daily et al. 1997; de Groot et al. 2002; Millennium Ecosystem Assessment, 2005). However, compared to economic and biophysical values, social, cultural, and other non-material values of ecosystems and biodiversity have generally been neglected in much of the literature on ecosystem services. In addition-although more recent research has made substantial progress in attempting to better integrate social perspectives and valuation techniques into the ecosystem services framework and to enable a more comprehensive representation of sociocultural values in theory and practice-social and cultural values can be difficult to measure and often require the use of approaches and methods that may include qualitative measures, constructed scales, and narratives, setting up as a kind of deliberative process involving the use of locally defined metrics.

As for the methods adopted to evaluate sociocultural ecosystem services in economic terms, there are several that measure their contribution to welfare; these fall into two categories: market and nonmarket. The former is based on real transactions and behaviors, and the latter engages users through surveys.

Market methods are in turn divided into direct and indirect.

In the former case we use the market price or production function, identifying the contribution made by the ecosystem service to another good, or the cost-based method that measures costs from production to supply.

In the second (stated preference method), reference is made to hedonic price and travel cost. Non- market methods, also called stated preference methods, are used when one intends to evaluate the impact of hypothetical changes in the ecosystem and the provision of its services. Among the most relevant are contingent valuation and choice experiments.

In addition to these two macrocategories, there are others that cannot be considered true economic evaluation tools because they use data collected from other studies: meta-analyses and value transfer. Cultural ecosystem services, for example, are almost always not monetizable, making the use of direct market valuations impossible. In some cases, it is possible to exploit indirect ones such as the cost of travel for recreational, tourism services and hedonistic pricing for aesthetic values.

Non-market methods are the most widely used because they allow the impact of a hypothetical physical transformation of the ecosystem on well-being to be measured, producing an ex-ante estimate that is useful for understanding possible scenarios and where action is needed.

The methods of the surveyed preferences are challenged, however, by the prevalence of users' opinions and needs over the maintenance or improvement of ecological conditions.

Below is an illustration of the main features of the different methods.

Indirect market methods

When cultural ecosystem services are not tradable, it is, in fact, necessary to use indirect market methods to measure the implicit price, through that of a good traded in the market.

Hedonistic valuation makes it possible to identify the impact that certain environmental conditions or ecosystem services have on the price of a good.

The large number of data needed, the risk of omitting important variables, and the willingness to pay for a specific area are the critical issues in this mode.

The cost of travel is another example. Typically used for tourist and recreational sites, it allows for estimating the value associated with their cultural ecosystem services. The expenses incurred for travel, including time, cost of transportation and access, reveal the willingness to pay to take advantage of specific services.

To be effective, a multiplicity of respondent information and surveys conducted at different times and seasons are required. Processing is done through statistical analysis and complex models. The difficulty to be addressed is the measurement of actual time spent, made greater in cases where the site to be visited is not the sole objective of the trip.

Non-market methods

Nonmarket methods are based on direct consumer surveys. Contingent valuation is constructed from the questionnaire that asks people how much they are willing to pay to maintain a certain ecosystem benefit and how much they would accept in return for giving it up.

Analysts must calculate the average willingness to pay and willingness to accept of respondents and multiply that value by the amount of people using the resource, thus estimating the total economic value attributed.

The construction of the survey is based on a six-step process:

1. definition of the problem to be evaluated and the hypothetical target market, selecting the ES and mechanisms to be used.
2. sample definition, deciding how many and which users to involve.

3. development of the basic structure of the questionnaire, choosing the appropriate form depending on t h e context and needs of the research.
4. identification of people's level of knowledge with respect to the asset, allows questions to be formulated and useful information to be presented appropriately.
5. Implementation of the questionnaire.
6. Analysis of results and estimation of mean value.

The process is suitable in any situation and can estimate both use and nonuse value.

The choice experiments method is based on the assumption that each good can be described in terms of characteristics and attributes. It asks respondents to indicate preference among several alternatives, allowing users to report in a simple and intuitive way the value they assign to benefits. It is developed in three main steps:

7. problem definition;
8. survey design;
9. Statistical analysis of data.

Other methods

Meta-analysis allows the synthesis and analysis of available empirical evidence on a given topic. Comparing and combining economic evaluations from different research allows for better estimates and to detect the factors that determine their variations. It also allows for generalization of results. Having identified the relevant literature, the data collected and the characteristics of the studies are coded and entered into a database. The accuracy of the results depends on the availability of the data and the statistical techniques through which they are processed.

Value transfer employs, at the site under analysis, economic assessments previously conducted in a similar area. It involves a simple approach or the use of value functions. The first approach is based on univariate transfer; averages of values from one site to another are reported directly. The second approach from the reference data constructs value functions for the site under consideration. Errors may occur in the original site estimate, during the transfer from one site to another, or in the calculation of the total economic value, compromising the results of the study.

3 The Evaluation Process According to MAES: Ecosystems, Indicators and Mapping

In this section, we will take as our main reference the Reports produced by MAES in the period 2013–2018, which are fundamental for the understanding and choice of the method to be applied for the elaboration of the Ecosystem Services assessment matrices.

The MAES is a working group established in 2012 by the European Commission, with the extended name of Mapping and Assessment of the Ecosystem and their Services, which has set itself the objective of supporting and increasing the mapping of ecosystems and their services for the European territory.

Within this project, four interim reports on specific aspects were published (2013, 2014, 2016a, 2016b) and the final report was published in 2018 by the European Union. In addition, a specific report on soil ecosystems was published in 2018.

In the first report 'Mapping and Assessment of Ecosystems and their Services. An analytical framework for ecosystem assessments under action 5 of the EU biodiversity strategy to 2020', from 2013, highlights the fact that mapping and analysis of ecosystems and ecosystem services is one of the key points of the EU Biodiversity Strategy.

In that document, it is noted that biodiversity plays a key role in the structural definition of ecosystems that are essential for maintaining basic ecosystem processes and sustaining ecosystem functions, the latter being defined as the capacity or potential to provide ecosystem services. Ecosystem services derive from ecosystem functions and through these goods and benefits are produced for humans; for this reason, ecosystem services are subject to economic valuation, although not all of them can be measured in monetary terms.

With reference to the latter, the importance of including other types of assessments, such as those on health, social conditions or conservation aspects, is highlighted. The report also states that healthy ecosystems (in good condition) have a full potential for ecosystem functions (consequently, ecosystem services) that are supported by biodiversity, for which the various aspects that contribute to ecosystem functioning are defined and which, in some cases, directly determine ecosystem services.

The second report, 'Mapping and Assessment of Ecosystems and their Services - Indicators for ecosystem assessments under Action 5 of the EU Biodiversity Strategy to 2020', from 2014, provides a list of reference indicators for measuring ecosystem services at the national scale. These indicators are defined for the three ecosystem services (provisioning, regulating and maintaining, cultural) and for each of the main ecosystems, i.e. forests, agro-ecosystems (crops and grasslands), running and still waters (lakes, rivers, groundwater and wetlands), marine and transitional areas.

The third report, 'Mapping and Assessment of Ecosystems and their Services - Mapping and assessing the condition of Europe's ecosystems: Progress and challenges', from 2016, concerns the mapping and analysis of ecosystems.

The fourth report, 'Mapping and Assessment of Ecosystems and their Services - Urban ecosystems', also published in 2016, addresses the topic of mapping and analysis of urban ecosystems (defined as socio-ecological systems that are composed of green infrastructure, understood as networks of urban green spaces, and built infrastructure) and proposes indicators to assess their conditions as well as that of urban ecosystem services.

This document is intended to support European policies concerning urban areas, represented by the EU Urban Agenda (2015), the EU Biodiversity Strategy (2011), the Green Infrastructure Strategy and Nature-based solutions (NBS).

Several case studies of European cities are presented in the report, including examples of the identification of urban ecosystems and/or ecosystem services and their representative indicators and mapping.

The fifth report, 'Mapping and Assessment of Ecosystems and their Services: An analytical framework for ecosystem condition', from 2018, identifies indicators for mapping and assessing the condition of ecosystems at the European level.

In particular, it provides a set of specific indicators for the assessment of ecosystem conditions for each ecosystem type, as well as a core set with key indicators to support an integrated ecosystem assessment per ecosystem type.

3.1 Classification of Ecosystems

With regard to the classification of ecosystems, the aforementioned fifth report, 'Mapping and Assessment of Ecosystems and their Services: An analytical framework for ecosystem condition', classifies ecosystems by correlating them with land uses and grouping them into six macro-families, which are described as follows

1. *Urban Ecosystems*

   An urban ecosystem is the ecological system located within a city. Like any other ecosystem, urban ecosystems are composed of interacting physical and biological components. Consequently, indicators measuring the condition of urban ecosystems and the pressures acting on them concern both the built and green infrastructure that together constitute cities.
2. Agroecosystems

   Agricultural land use is the main land use in the European Union and accounts for 45% of its total area. The type of ecosystem addressed in this section is agroecosystems, understood as communities of plants and animals that interact with their physical and chemical environment and have been modified by humans to produce food, fibre, fuel and other products for human consumption and processing.
3. Forests and Woodlands

   Forests and woodlands cover about 40 per cent of the EU's land area and are home to much of Europe's biodiversity. Similarly, forests provide multiple ecosystem services that support and satisfy human needs. As a result, society benefits from forest services and at the same time modifies forest ecosystems through a range of direct and indirect pressures, e.g. land use, climate change, air pollution and invasive alien species contribute to shaping the condition of forests.
4. Natural Ecosystems (Heaths and Shrublands, Sparsely Vegetated Land and Wetlands)

   This section focuses on ecosystems that are largely covered by the Habitats Directive (HD) and the Birds Directive (BD), the so-called Nature Directives because of their high biodiversity values. According to the MAES typology, these ecosystems are "Heaths and Shrublands", "Sparsely Vegetated Land" and "Wetlands".
5. Freshwater Ecosystems

   Freshwater ecosystems include rivers, lakes and groundwater. Their condition and functioning are closely linked to natural ecosystems at the water-land interface, such as riparian areas, floodplains and wetlands. This section of land use types focuses on rivers, lakes and groundwater, while wetlands are covered in Sect. 4 and transitional and coastal waters in Sect. 6.
6. Marine Ecosystems

   This section focuses on marine ecosystems, defined as encompassing all marine waters, including those at the land/sea interface with salinities above 0.5‰. Following the MAES typology, four ecosystems are considered: marine inlets and transitional waters, coastal waters, shelf waters and the open ocean.

3.2 Ecosystem Service Indicators

The report also identifies, in illustrative tables, the soil ecosystem services related to the three main categories, identifying the appropriate processes required to produce these services, and provides, again in tables, for each ecosystem service, indicating the relationship with the CICES classes and types, a list of related indicators, for which the reference spatial dimension (regional or local) of application is defined, the nature: supply or use services, and the availability of data.

The document also describes indicators for the assessment of pressures and soil ecosystem conditions. With regard to pressures, reference is made to: agricultural land management intensity; forest management intensity; soil sealing; soil disturbance; land use change; soil moisture; contaminated sites; gross nutrient balance; water extraction; organic matter loss; soil erosion. Soil ecosystem conditions refer to: soil erosion; soil sealing; soil contamination; water supply capacity; soil nutrient concentration; soil organic carbon; soil moisture; bulk density; soil biodiversity.

The document also contains a summary of the impacts that each pressure causes on biodiversity in relation to different ecosystems (urban, agricultural, grassland, forest, shrub, wetland, inland aquatic, marine). The pressures considered are the following: habitat modification, climate change, overexploitation, invasive alien species, pollutants and nutrient surplus.

The assessment of impacts is represented with classes referring to intensity (low, moderate, high, very high) and with indications of expected trends (decrease, maintenance, increase, very rapid increase).

3.3 Mapping Ecosystem Services

Ecosystem services, due to their peculiarities, are localised in specific areas and it is therefore possible to map them spatially; several studies have highlighted how they can be considered "spatially specific" (Boyd and Banzhaf, 2006) or "spatially explicit" (Haase et al. 2014) and this is the characteristic that makes them a non-tradeable and non-saleable good. With regard to these considerations, it is noted that ecosystem services refer to a particular spatio-temporal dimension, in the sense that they can be used in place but also in distant places, as some of them (e.g. water for drinking) can be formed in a particular place and period but be used in other places in later periods.

The forms of mapping and representation, as well as the methods of evaluation, vary and, to some extent, depend on multiple factors, such as: the objective of the analysis, the spatial scale of restitution, the information available, the level of depth of the investigation, the resources that can be used and the tools - especially IT tools - available. There are numerous studies and proposals in this regard (Burkhard and Maes, 2017; Crossman et al. 2013; Hauck et al. 2013; Martínez-Harms and Balvanera, 2012) and there is no single shared procedural model, although the indications of the MAES project advocate the use of models capable of ensuring, at the very least, the scientific accuracy, reproducibility of the methods and reliability of the analyses produced (scientific accuracy, reproducibility, credibility).

The mapping, considering the two main forms of restitution suggested by the 2016 MAES report, can represent the supply or provision of ecosystem services (supply), which is indispensable for making assessments, or the demand of the population. The latter must be such as to allow cross-referencing with the former, both in order to verify the territorial coincidence or otherwise with supply and to direct spatial and urban planning; this in view of the fact that land-use transformations decisively condition the supply of ecosystem services and that, therefore, appropriate choices of urban plans and projects can ensure its preservation or enhancement.

In fact, the mapping of ecosystem services, combined with the representation of their value (qualitative or quantitative, not necessarily monetary), has a strong communicative potential towards decision-makers and towards local actors and the non-expert population. In this regard, it should be noted that for the construction of the maps, particularly those of demand, it may be appropriate to follow paths of involvement of the various actors, who will certainly be able to provide their input with regard to expectations and needs, but also to criticalities, due to the absence or lack of services, and priorities for action.

Another relevant aspect is the choice of ecosystem services to be considered for mapping purposes. These may vary in number and typology, depending on the characteristics and size of the territorial context to be analysed and also on the final purpose; in general, the selection should be conducted by taking as a reference the ecosystem services already identified in schemes recognised at European level and taking into account the aggregation of services into the main categories (supply, regulation, cultural and, for some schemes, habitat), so that all of them can be represented.

In this regard, with regard to the different level of depth when mapping ecosystem services, some considerations from the MAES Reports are echoed.

In the second MAES report, 'Mapping and Assessment of Ecosystems and their Services - Indicators for ecosystem assessments under Action 5 of the EU Biodiversity Strategy to 2020', from 2014, the topic of mapping ecosystem services is considered and three levels of depth and representation are outlined.

At the first level, representing a proxy of a given ecosystem service, restitution is based on the use of indicators that can be explicitly derived from land use and land cover (e.g. Corine Land Cover) or biodiversity monitoring or forest inventory. The document emphasises that some of the indicators defined by MAES are areal elements or, in any case, spatially permeable and can, therefore, be used to construct maps.

At a second level, there is the use of a set of indicators to be composed of each other so that ecosystem services can be assessed; in this case, land use data are combined with information that takes into account the relationship between uses and ecosystem services and allows quantitative measurement for different locations or aggregations at different scales.

The third level requires the use of biophysical process modelling in a GIS environment or other software capable of cross-referencing information from several indicators. In this case, open-access programmes such as InVEST and ARIES can be used.

In the third MAES Report, 'Mapping and Assessment of Ecosystems and their Services - Mapping and assessing the condition of Europe's ecosystems: Progress and challenges' (2016), the mapping of ecosystem services is defined as the identification and delineation of the spatial extent of different ecosystems through the spatial integration of a wide range of data on land/sea cover and environmental characteristics. It defines ecosystem services capacity mapping as the combination of ecosystem maps with data on the conditions or relationships between conditions, functions and capacities of ecosystem services.

With regard to the ecosystems to be mapped, the types identified are the following: urban, agricultural-cultivated, grassland, forest, shrub and dryland, sparse vegetation, wetland, river and lake, marine inlet and transitional waters, coastal, marine shelf, and open ocean. The document provides a correlation table between MAES level 2 and level 1 or 2 of the EUNIS habitat classification system, which in turn can be associated with the Corine Land Cover categories.

Regarding modalities, the report points out that several methods can be used to map the supply of ecosystem services. These range from the use of indicators such as carbon storage capacity, crop yield or water infiltration into the soil (Layke et al. 2012), to the method based on expert opinion on the performance of different types of land cover to generate specific services, to that using process models. The paper notes that indicators for mapping the supply of ecosystem services have already been developed, citing for example authors Burkhard et al. (2012), Haines-Young et al. (2012).

With regard to the methods of cartographic restitution of ecosystem services at a territorial scale, some significant experiences are recalled.

In the article by B. Burkhard et al. from 2009, an initial study phase is proposed and defined as suitable for the development of integrated strategies for the provision of ecosystem services at territorial level.

This phase is based on the analysis of Corine Land Cover data that allow for the expert assessment of the different potential of different land covers to provide ecosystem services and, therefore, for their restitution by means of cartographic representations; the authors emphasise that, in particular, the spatial explication of information obtained through mapping has a high potential in favouring the understanding of complex and interacting systems (Dresner, 2008).

For the elaboration of the maps, a matrix is developed, through which the different capacities of land cover classes to provide ecosystem services are identified; in detail, the 44 Corine Land Cover classes are crossed with the 29 ecosystem services selected by taking into consideration the lists proposed by De Groot (2006), the Millennium Ecosystem Assessment (2005) and Costanza et al. (1997), as well as the list on the integrity of ecological components described by Müller and Burkhard (2007). Each crossroad is assigned a rating class defined according to the following capacity scale: 0 = not relevant; 1 = not very relevant; 2 = relevant; 3 = medium relevance; 4 = high relevance; 5 = very high relevance.

The allocation of these classes is based on an initial expert assessment and feedback from different case studies. It can be tested by applying it to other cases - with additions, modifications and adaptations - in relation to available information and models from expert estimates, as well as by assessing the importance of particular ecosystem services in the territorial contexts analysed.

The methodology was applied in the case study of the Halle-Leipzig area, a region in the central-eastern part of Germany, within the framework of the PLUREL project; in this case study, two maps were produced to restore the spatial distribution of ecosystem services, as of 1990 and 2000, in order to be able to make a comparison and to verify and highlight the changes resulting from changes in land cover.

Another application, at a more recent date, was carried out in the Baltic geographical area and the results are described in the publication edited by Kalvane I., Burkhard B.,

Ruskule A., Bojars E., "Methodological Guidelines for Mapping and Assessmente of Grassland Ecosystem Service", Baltic Environmental Forum - Latvia (2014).

This document provides guidance for the mapping of ecosystem services and their assessment at the local level. In the introduction, it is noted that the quality of ecosystem service mapping studies depends on the methods used to quantify and map, the information available and the purpose of use.

The study applies the crossing matrix between the Corine Land Cover categories and the selected ecosystem services, with each crossing assigned a relevance value from 0 (absent) to 5 (highest).

This procedure is based on expert estimation, biophysical quantification and the result of empirical modelling, estimating the capacity to provide ecosystem services that are attributed to land use or land cover classes.

The paper points out that for more complex analyses, GIS processing models are available for the assessment of ecosystem services, and briefly recalls and describes those that are most widely applied, namely InVEST, ARIES and SolVES.

4 The Valuation of Ecosystem Services Within the Teach4You Research Project "Technologies for Adapting to Climate Change and Improving Quality of Life".

On the basis of the above, in the context of the Teach4You Research Project "Technologies for adapting to climate change and improving quality of life", it was decided to refer to the method of estimating the capacity to provide Ecosystem Services in relation to different types of land use.

The classes of Ecosystem Services considered are those based on the lists proposed by: Costanza et al. (1997), de Groot et al. (2002), Millennium Assessment (2005) and de Groot (2006). In these lists, Ecosystem Services are grouped into four main categories: regulatory services, provisioning services, habitat services and cultural services.

With regard to the land use classes considered - grouped according to the six macro categories proposed by MAES and outlined above - these refer to the Copernicus database.

The Copernicus Land Monitoring Service (CLMS) provides geographical information on land cover and related changes, land use, vegetation status, water cycle and land surface energy variables. It enables applications in various sectors such as urban and spatial planning, forest management, water management, agriculture and food security, conservation and restoration of the natural environment, rural development, ecosystem accounting and climate change mitigation/adaptation.

In particular, as far as it is of research interest, the land monitoring service offered by Copernicus covers:

- Information on land cover and land cover change in European countries over the period 2000–2018.

  This information on land use types and their spatialization is taken from the CORINE Land Cover programme. The CORINE programme was initiated in the European Union in 1985 (EEA, 1994) and the database includes 44 land cover classes in Europe. The Corine Land Cover (CLC) project was created specifically to survey and monitor land cover and land use characteristics, with particular attention to environmental protection requirements. The initiative, co-financed by the Member States and the European Commission, was joined in 2000 by 33 countries, including Italy, where the National Authority for managing the project was identified in APAT, as the national focal point of the European EIOnet network.

– Information on land cover and land cover change in European cities in the period 2012–2018.

  This information on land use types in urban areas and their spatialization is taken from the Uban Atlas programme. The Urban Atlas (UA) dataset provides land cover/land use data on Functional Urban Areas (FUAs). The main datasets are: Urban Atlas 2006, Urban Atlas 2012, Urban Atlas 2018, the change datasets 2006–2012 and 2012–2018 and the Street Tree Layer 2012 and Street Tree Layer 2018 products. At the European level, the 2006 dataset includes 319 FUAs representing areas with more than 100,000 inhabitants, the 2012 dataset includes 785 FUAs with more than 50,000 inhabitants, and the 2018 dataset consists of 788 FUAs.

  From the production of the 2012 dataset onwards, the nomenclature identifies 17 urban classes with a minimum map unit of 0.25 ha and 10 rural classes with a minimum map unit of 1 ha. The update frequency is six years.

5 Conclusions

In closing this brief discussion, it is worth pointing out how the assessment of ecosystem services is a crucial element in the planning process of territories and cities, especially from a point of view that places at the canter the creation of a different scientific and cultural approach that contributes to creating a collective knowledge and awareness of the real value that they assume in future development choices aimed at guaranteeing human wellbeing, since they constitute an inseparable and bidirectional link between man and ecological systems.

Designing according to nature-based solutions may, in fact, constitute an innovative way, as it is based on a vision that places nature at the center of transformation choices, also through the identification of alternative scenarios that are both efficient and economically appropriate.

Evaluating ecosystem services becomes, therefore, a basic action to prepare design scenarios for cities and territories by estimating ex ante policies, projects and realization of works, and works and, furthermore, to be able to measure the level of wellbeing of the population and prefigure possible future evolutions, providing decision-makers with the tools to mitigate the possible effects of actions that could affect wellbeing and quality of life.

Acknowledgements. This research work is the result of the research activities conducted within the following PNRR research project funded by the European Union – NextGenerationEU: "Pilot Project 4.6.1" "Climate adaptation plans for the reduction of the ecological footprint and ecological debt, aimed at improving the conservation and transformation response in terms of resilience and quality of life in the Calabrian and Lucanian urban and territorial systems" (Goal 4.6 - Planning for Climate Change to boost cultural and natural heritage: demand-oriented ecosystem services based on enabling ICT and AI technologies - Tech4You Innovation Ecosystem), Finanziato dall'Unione Europea - Next Generation EU, Missione 4 Componente 2 - CUP C33C22000290006 (ECS_00000009).

References

Arcidiacono, A., Ronchi, S., Salata, S.: Computational Science and its Applications – ICCSA **9157**, 35–49 (2015). https://doi.org/10.1007/978-3-319-21470-2

Bateman, I.J., et al.: Economic Valuation with Stated Preference Techniques: A Manual. Edward Elgar Publishing Ltd., Cheltenham (2002). https://www.cabdirect.org/cabdirect/abstract/20043091507

Boyd, J.W., Banzhaf, H.S.: What are Ecosystem Services? The Need for Standardized Environmental Accounting Units (January 2006). Resources for the Future, Discussion Paper No. RFF DP 06–02. SSRN: https://ssrn.com/abstract=892425 or https://doi.org/10.2139/ssrn.892425

Burkhard, B., Kroll, F., Müller, F., Windhorst, W.: Landscapes' capacities to provide ecosystem services - a concept for land-cover based assessments. Landscape Online **15**(1), 1–22 (2009)

Burkhard, B., Kroll, F., Nedkov, S., Müller, F.: Mapping ecosystem service supply, demand and budgets. Ecol. Ind. **21**, 17–29 (2012). https://doi.org/10.1016/j.ecolind.2011.06.019

Burkhard, B., et al.: Land cover-based ecosystem service assessment of irrigated rice cropping systems in southeast Asia-An explorative study. Ecosystem Services **14**, 76–87 (2015). https://doi.org/10.1016/j.ecoser.2015.05.005

Burkhard, B., Maes, J.: Mapping ecosystem services. Advanced Books (2017). https://doi.org/10.3897/ab.e12837

Costanza, R., et al.: The value of the world's ecosystem services and natural capital. Nature **15**(387), 253–260 (1997)

Crossman, N.D., et al.: A blueprint for mapping and modelling ecosystem services. Ecosyst. Serv. **4**, 4–14 (2013)

De Groot, R.S., Wilson, M.A., Boumans, R.M.J.: A typology for the classification, description and valuation of ecosystem functions, goods and services, special issue: the dynamics and value of ecosystem services: integrating economic and ecological perspectives. Ecological Economics **41**, 393–408 (2002)

De Groot, R.S.: Function-analysis and valuation as a tool to assess land use con- flicts in planning for sustainable, multi-functional landscapes. Landsc. Urban Plan. **75**(3–4), 175–186 (2006). https://doi.org/10.1016/j.landurbplan.2005.02.016

De Groot, R.S., Alkemade, R., Braat, L., Hein, L., Willemen, L.: Challenges in integrating the concept of ecosystem services and values in landscape planning, management and decision making. Ecol. Complex. **7**(3), 260–272 (2010). https://doi.org/10.1016/j.ecocom.2009.10.006

Dresner, S.: The Principles of Sustainability, 2nd edn. EarthScan, Oxford (2008)

Haase, D., et al.: A quantitative review of urban ecosystem service assessments: concepts, models, and implementation. Ambio **43**, 413–433 (2014). https://doi.org/10.1007/s13280-014-0504-0

Haines-Young, R., Potschin, M., Kienast, F.: Indicators of ecosystem service potential at European scales: mapping marginal changes and trade-offs. Ecol. Ind. **21**, 39–53 (2012)

Hauck, J., Görg, C., Varjopuro, R., Ratamäki, O., Jax, K.: Benefits and limitations of the ecosystem services concept ienvironmental policy and decision making: some stakeholder perspectives. Environmental Science and Policy **25**, 13–21 (2013)

Hostetler, M., Allen, W., Meurk, C.: Conserving urban biodiversity? creating green infrastructure is only the first step. Landsc. Urban Plan. **100**(4), 369–371 (2011). https://doi.org/10.1016/J.LANDURBPLAN.2011.01.011

Kalvane, I., Burkhard, B., Ruskule, A., Bojars, E.: Methodological Guidelines for Mapping and Assessmente of Grassland Ecosystem Service, Baltic Envirnmental Forum – Latvia (2014)

Layke, C., Mapendembe, A., Brown, C., Walpole, M., Winn, J.: Indicators from the global and sub-global millennium ecosystem assessments: an analysis and next steps. Ecol. Indic. **17**, 77–87 (2012)

Lovell, S.T.: Multifunctional urban agriculture for sustainable land use planning in the United States. Sustainability **2**(8), 2499–2522 (2010). https://doi.org/10.3390/su2082499

MA (Millennium Ecosystem Assessment). Ecosystems and Human Well-Being: Synthesis. Island Press, Washington, DC (2005a)

MA (Millennium Ecosystem Assessment). Synthesis Report. Strengthening capacity to manage ecosystems sustainably for human wellbeing. Millennium Ecosystem Assessment Offices, p. 219 (2005b)

Maes, J., et al.: Mapping and Assessment of Ecosystems and their Services. An analytical framework for ecosystem assessments under action 5 of the EU biodiversity strategy to 2020. Publications office of the European Union, Luxembourg (2013)

Maes, J., et al.: Mapping and Assessment of Ecosystems and their Services. Indicators for ecosystem assessments under Action 5 of the EU Biodiversity Stategy to 2020. Publications office of the European Union, Luxembourg (2014)

Maes, J., et al.: Mapping and Assessment of Ecosystems and their Services. Publications office of the European Union, Luxembourg, Urban ecosystem (2016)

Maes, J., et al.: Mapping and Assessment of Ecosystems and their Services: An Analytical Framework for Ecosystem Condition. Publications office of the European Union, Luxembourg (2018)

Martínez-Harms, M.J., Balvanera, P.: Methods for mapping ecosystem service supply: a review. International Journal of Biodiversity Science, Ecosystem Services & Management **8**, 17–25 (2012). https://doi.org/10.1080/21513732.2012.663792

Mitchell, M.G.E., et al.: Reframing landscape fragmentation's effects on ecosystem services. Trends in Ecology & Evolution **30**(4), 190–198. https://doi.org/10.1016/J.TREE.2015.01.011 (2015)

Millennium Ecosystem Assessment (MEA), Ecosystem and Human Wellbeing: A Framework for Assessment. Island Press (2005)

Müller, F.: Indicating ecosystem and landscape organization. Ecological Indicators **5**(4), 280–294 (2005)

Müller, F., Burkhard, B.: An ecosystem-based framework to link landscape structures, functions and services. In: Mander, Ü., Wiggering, H., Helming, K. (Eds.): Multifunctional Land Use – Meeting Future Demands for Landscape Goods and Services. Springer, pp. 37–64 (2007)

Nedkov, S.: Flood regulating ecosystem services—mapping supply and demand, in the Etropole municipality, Bulgaria. Ecol. Ind. **21**, 67–79 (2012). https://doi.org/10.1016/J.ECOLIND.2011.06.022

Pielke, R.A.: The influence of land-use change and landscape dynamics on the climate system: Relevance to climate-change policy beyond the radiative effects of greenhouse gases. philosophical transactions of the Royal Society of London. Series A **360**, 1–15 (2002)

Santolini, R.: Paesaggio e sostenibilità: i servizi ecosistemici come nuova chiave di lettura della qualità del sistema d'area vasta. In: Riconquistare il Paesaggio, la Convenzione Europea del Paesaggio e la conservazione della biodiversità in Italia, MIUR - WWF Italia, pp. 232–244 (2008)

Santolini, R., Morri, E.: *Criteri ecologici per l'introduzione di sistemi di valutazione e remunerazione dei Servizi Ecosistemici (SE) nella progettazione e pianificazione.* In: La dimensione europea del consumo di suolo e le politiche nazionali, CRCS Rapporto 2017: 149–154, INU ed., Roma (2017a)

Santolini, R., Morri, E.: *Valutazione e mappatura dei Servizi Ecosistemici: strumenti di governance sostenibile del paesaggio.* Urbanistica 158, INU ed., Roma (2017b)

Stürck, J., Poortinga, A., Verburg, P.H.: Mapping ecosystem services: the supply and demand of flood regulation services in Europe. Ecol. Ind. **38**, 198–211 (2014). https://doi.org/10.1016/j.ecolind.2013.11.010
van der Meulen, S., et al.: Mapping and Assessment of Ecosystems and their Services. Soil Ecosystems, European Commission DG Environment, Bruxelles (2018)

The Development of the City Through the Green Strategy. The Use of Sustainability Indicators

Domenico Passarelli(✉)

Università degli Studi Mediterranea, 89124 Reggio Calabria, Italy
domenico.passarelli@unirc.it

Abstract. This research concerns the complex analysis of ecosystem relations in territorial areas and has resulted in the identification and development of sustainability indicators with the attempt to promote actions aimed at the protection and valorisation of natural, cultural and identity heritage of the territory. The ultimate aim is to build scenarios and predictive models for monitoring the impact of transformations. The inspiring principle focuses on the growing awareness that natural ecosystems must be maintained in dynamic balances to ensure viable future growth through the provision of goods and services based on the logic of ecosystem services. The research work has as its objective the improvement of the quality of life and includes aspects relating to the environmental, social and cultural well-being of citizens linked to a better use of the categories of ecosystem services (life support, supply, regulation of cultural lore). In particular, from an infra- and intra-generational ecological perspective, a strengthening of identity values is expected, of the usability of the natural and cultural heritage of smaller urban centers and internal areas, of the quality of green and blue infrastructures aimed at sustainable use of territorial and environmental resources. The expected results are measured on the basis of the key indicators established by the European Commission to monitor the diffusion of innovation at a regional level.

Keywords: strategy · indicators · sustainability

1 Understanding the Phenomena at Play in Relation to the Objectives to Be Pursued

1.1 Knowing to Plan

Ecosystem services (goods, resources and services) are essential for survival.

Ecosystem services are the series of services that natural systems generate for the benefit of man: according to the definition proposed by the MEA - Millennium Ecosystem Assessment, ecosystem services are the "multiple benefits provided by ecosystems to mankind" (MIA, 2005). Ecosystem Services can be grouped into 3 main categories: such as: SE regulation of atmospheric gases, climate, water, erosion, prevention of hydrogeological instability, regulation of pollination, ha-bitat for biodiversity; SE of food supply, raw materials, fresh water, biological variability; cultural SEs, such as aesthetic, recreational, educational, spiritual, artistic and identity values. The importance

F. Moraci et al. (Eds.): NMP 2024, LNNS 1272, pp. 80–90, 2025.
https://doi.org/10.1007/978-3-031-82927-7_5

of ecosystem services is therefore very high as they, directly or transmit, influence and support human life and well-being in terms of health, access to primary resources, sustenance…etc. Numerous studies confirm that our consumption and production systems are simply unsustainable. If current trends continue, regardless of country and income level, future generations will face a situation characterized by more extreme temperatures and weather events, reduced biodiversity, greater resource scarcity and higher levels of pollution. People cannot live well if the environment and the economy are in poor condition. Yet, despite the magnitude of the task ahead of us, it is still possible to build a sustainable future. The time has come to measure ourselves with new parameters aimed at the quality of the place, the well-being of the community, also through new governance. It is worth remembering that the impact of the research project on improving the quality of life includes aspects relating to the environmental, social and cultural well-being of citizens linked to a better use of the categories of ecosystem services (life support, supply, regulation of cultural values). In particular, from an infra- and intra-generational ecological perspective, a strengthening of identity values is expected, of the usability of the natural and cultural heritage of smaller urban centers and internal areas, of the quality of green and blue infrastructures aimed at sustainability of territorial and environmental resources.

1.2 Land Consumption and Ecosystem Services: An Indissoluble Relationship

Ecosystem services are considered one of the important issues for the sustainable development of the territory. Nonetheless, even today there is no full awareness of the need to save soil and even more so there is less conviction of the importance of ecosystem services in relation to their use in urban planning choices. A very interesting case study is that of the Municipality of Bruino (province of Turin), a pilot experience conducted as part of the drafting of the structural variant to the PRG [1]. Unfortunately, no value has been given to the soil and it is difficult to recognize it as a common good: it is a non-renewable resource and is threatened by growing both anthropic and natural pressures: contamination due to the gutting and burial of polluting substances harmful to health; desertification and hydrogeological instability and others which are irreversibly compromising the functionality of the soil resource. Appreciation must be given to the bill "containment of land consumption and reuse of built land n° 2383 which has set among its objectives that of valorising agricultural areas", promoting and protecting the agricultural activity of the landscape and the environment, in order to prevent the soil, as an essential resource for the balance of ecosystems, from being excessively eroded, waterproofed and consumed by urbanization with highly impactful and negative consequences in social, environmental and business terms. The increase in built land also reduces the ability of the environment to absorb CO2 and therefore to counteract the greenhouse effect and reduce global warming and its harmful consequences. Obviously, lower land consumption also has very significant economic and social consequences. A territory with settlements dispersed like a leopard is much more expensive and difficult to infrastructure and manage, while the right level of concentration allows for more economical management as well as the achievement of threshold levels not only for the cost-effectiveness of technological and transport networks but also for the creation of sustainable and identifiable social networks. The key principle is that the soil resource is a vital, limited and non-renounceable resource and consequently a common commitment is needed to contain it, according to the general objectives set by the EU and in coherence with other national initiatives. It should be underlined that the practice of reusing already built land and urban regeneration represent the key tools for limiting land consumption and therefore the same cannot be framed in a sectoral perspective but must be addressed in a unitary vision of territorial policy and integrated ecology. The DDL on urban regeneration no. 1131 in art. Seems to be going in this direction. 2 paragraph 1, provides for an area identified by the municipalities called "green belt" with agricultural, ecological-environmental and recreational functions, consistent with the conservation of ecosystems pursuant to article 6 of law 14 January 2013, n. 10, aimed at preventing land consumption and encouraging the absorption of carbon dioxide emissions from the atmosphere through the increase and valorisation of the tree heritage, energy efficiency, the absorption of fine particles, as well as reducing the "heat island" effect, while promoting regular rainwater collection. But it's not enough. We need a law to limit land consumption. Although in recent years the extent of the processes of anthropisation and soil waterproofing has been reduced, compared to the quantities measured in the first decade of the 2000s when the average amount of land consumed exceeded 70 hectares per day (ISPRA, The consumption of soil in Italy, 2014; CRCS Land Consumption Research Center, National Report, 2010) land consumption continues today with an average rate of around 15 hectares per day. A law is necessary to reaffirm the non-conforming nature of territorial structural level plans and free planning choices from the "weight" of the "plan residue"; that is, the transformation and new urban development forecasts contained in the general regulatory plans in force, which are difficult to annul except with the risk of long and expensive appeals for the Administrations. A law that establishes a forfeiture of urban planning provisions if not implemented within a period of time, similarly to what happens for public constraints which cease to be valid after 5 years from their affixing without implementation.

1.3 The Urban Bioregion: A Possible Experimentation

This research operates in a new relational dimension, the urban bioregion: only sustainability-oriented planning and therefore a new form of plan will be able to generate a sustainable development process both from an environmental and socio-economic point of view and within the renewed urban planning instrument, ecosystem services find space in a new territorial dimension. Bioregionalism is an ethical, political, ideological approach, linked to the territory in which one lives, considered as a homogeneous whole from the morphological point of view and that of living beings. The term bioregion comes from the Greek word bios (life) and the Latin regere (to govern). It is therefore a question of considering a homogeneous geographical territory in which the rules dictated by nature should be predominant and not the laws that man would have defined. "The government of nature", this is how Kirkpatrick Sale defined the deepest meaning of bioregionalism. The bioregion is a territorial unit, with homogeneous physical and ecological characteristics. Since there is no standard size, we could consider it a synthesis between a biogeographical district and the territory of a province. Although bioregions are all interrelated, each person lives within a specific and determined bioregion. Peter Berg, one of the founding fathers of bioregionalism, defined the bioregion as "both the geographical terrain and the terrain of consciousness", understood as the consciousness of places on the part of the inhabitants! Bioregionalism is therefore that "form of decentralized human organization which, aiming to maintain the integrity of biological processes, life formations and geographical formations specific to the bioregion, helps the material and spiritual development of human communities that inhabit it" (Thomas Rebb, 1998). In fact, once one has recognized one's own bioregion, one's own "place" be it urban, rural or wild, one must live there entirely, think in a bioregional way, which is not the adhesion to a new static ideology but the discovery, and daily practice, of a new personal and ecological way of living in harmony with nature (Gary Snyder's "real work", 1980) [2]. The elaboration of this concept is the responsibility of the Canadian intellectual Alan Van Newkirk who, studying human geography, came to the conclusion that communities of living beings interact with each other and with their physical environment, by organizing themselves into wholes that show continuity between physical and ecological characteristics. The determination of actions and projects aimed at redeveloping the settlement and naturalistic-environmental systems that compose them requires the identification of the main problems relating to these contexts and the possible directions of eco-sustainable development already undertaken or still to be undertaken. These themes are receiving growing attention and a different level of in-depth analysis. Current urbanization processes have a heavy impact on human settlement and the living environment in general. Social disintegration and economic fragmentation are accompanied by the compromise of ecosystems, the loss of fertile soil and worrying phenomena relating to climate change. Based on the bio-regionalist approach, this contribution exposes the main references for the definition of a new territorial dimension to methodologically address (and with a utopistically concrete vision) the indicated problems. In this framework, the main point of reference is the idea of the territory conceived as a "common good" and the recovery of the co-evolutionary relationship between human settlement and ecosystem resources. In particular, the need to support a process of construction and recovery from the bottom of "place awareness" on the part of the inhabitants is indicated, as a key element for fueling conviviality of life and sustainable use of the territorial heritage, also suitable for generating self-sustaining local and regional economies. -sustainable. The following considerations are based in particular on the figure of the urban bioregion and its constituent materials: contextual knowledge and skills, ecosystem quality and hydro-geomorphological stability, polycentric settlement systems and regeneration/centrality of public space, self-sustainable local economies, mixed local energy systems, multifunctionality of agro-forestry areas and, last but not least, the redefinition in participatory terms of local bodies of political/administrative life.

2 Innovate the Rules. From the Traditional System to the Performance One

A renewed approach to territorial planning that includes energy efficiency and renewable sources, sustainable agriculture, soft mobility and accessibility, maintenance and safety of buildings, social regeneration of abandoned and degraded areas offers prospects for qualitative development of the territory. To translate it into concrete proposals it is necessary to measure ourselves with the specific characteristics of each territory, both in terms of the available resources, and in terms of the needs to be filled, and in terms of the social composition, i.e. the actors of the communities who live there but also and above all to interpret current needs. Each territory has different resources and potential and therefore a single project cannot be proposed and cannot be replicated everywhere. The main objective remains to propose coherent and compatible scenarios with respect to the endogenous resources present and the activation of self-propelled development mechanisms and the correct application of ecosystem services through nature-based solutions, recognizing those limiting factors that create difficulties of implementation in current practices. It is a challenge that requires responsibility and planning. The tools chosen are those useful for founding a widespread culture of an ecological and supportive nature: research and development, training and information. The European experience that best suits our level of study is the French one centered on the "schéma diréctéur", a term that designates, with general value, the project management tool also in fields other than urban and territorial planning. The direct scheme combines several municipalities and can be assimilated to a strategic plan at an inter-municipal level. In this context, the performance criteria of the settled communities must be fully accommodated. The performance interpretation recalls the idea of Kevin Lynch [3] who delves into the theme of the shape of the city according to a perceptive reading modality, as already addressed by the author twenty years earlier in the text The image of the city in which he places particular attention to the human scale by investigating the perception of spaces by the people who use them. Lynch, convinced that the planner's main concern should be to understand the physical environment and help shape it to meet the needs of citizens, starting from the analysis of existing forms to determine their effectiveness with respect to the initial objectives, formulated together with Lloyd Rodwin an innovative investigation system, useful for strengthening even the most consolidated planning methodologies. This system attributed particular importance to physical space, to the complex and dense relationships that flows establish with it and to the spatial structure of urban functions according to the concept defined as grain, or "the grain, i.e. the internal structure of a settlement, a fundamental aspect of its fabric, an aspect that is often confused with density. This term refers to the way in which the different elements that make up a settlement are mixed together in space. These elements can be: activities, building types, people or other aspects" [3]. The methodological path just outlined implies a "normative theory" based on the clarification of performance rules with an open method "whose physical forms must correspond to specific requirements, relating primarily to the biological characteristics of man and the morphological characteristics of the site" [4]. This interpretation is characterized by the construction of the city "providing a corpus of rules that not only constitute a criterion of analysis and measurability of the compatibility of the intervention on the new, but also a method of evaluation for the existing city" [5]. In conclusion, it is important to pursue the transition from traditional zoning rules (the so-called exigent rules) to performance rules, according to which public objectives and an updated regulatory framework are defined which allows the subjects interested in planning, and last but not least the citizens and cultural associations, to find an operational agreement capable of synthesising the different interests and needs such as to promote a correct approach to ecosystem services and more effective planning solutions to face the emerging challenges of the contemporary city.

2.1 The Development of the City Through the Green Strategy

A green strategy for the territory involved in the Tech 4 For You research project can only start from the recognition of the important role played by greenery in urban settlements and other sectors of the economy and society. The green deal for the south is the opportunity for a major green restructuring work for the territory, favoring the mitigation of seismic, hydrogeological and other environmental risks that limit and very often strongly inhibit the development of good planning practices. In light of this, quality job opportunities are needed, generated by companies capable of competing, growing and innovating. All this is possible if virtuous processes of promotion and strengthening of networks between business and research are triggered, through the construction of a modern and competitive technology transfer fabric in compliance with sustainability. The following considerations refer to a principle according to which the city is first and foremost a complex of networks: "cities are, by definition, plural, public and productive, they are the Petri dishes of experimentation and are guided in their evolution by society itself, except in exceptional cases such as Brasilia, created on the basis of a vision imposed from above [6]. In this context, green infrastructures will have to take on a multi-functional character, reconnecting the concept of greenways with that of ecological networks with the aim of bringing advantages to the settled communities but at the same time enhancing the ecological aspects present in a sustainable development perspective. In this sense, green networks take on a different role than in the past in order to increase the degree of biological diversity and self-propelled and regenerative capabilities. In the context of an emerging consciousness and in the face of the current environmental imbalances of the contemporary city, the idea of a green city is taking shape, that is, a renaturalization of the city through real initiatives of structural integration of greenery with the built environment (creation of urban gardens and wooded areas, habitats for wildlife, ponds and wetlands and natural and artificial plant corridors, where the horizontal space does not allow the insertion of further and appropriate spaces greens). We are now far from considering greenery as a simple merely decorative fact, especially since it can significantly contribute to guaranteeing a high quality of living within an ecological vision of the city. The quality of the territory necessarily also passes through both public and private urban greenery. Greenery, as one of the most important elements of attraction, also becomes a factor of competitiveness for the city's economy, of quality for the life of its citizens, of the city's landscape identity, it becomes a "future common good" [7]. Urban spaces are made up of public and private greenery, in their various typologies, from the small garden to the large park. To these spaces must be added the tree-lined avenues which give the whole the shape of a system. Some thematic indications on renaturalization offered by the Urban Agenda (goals 6–11-13–15) invite us to plant evergreen trees resistant to climatic stress, to the use of filter-plants as a bioremediation strategy, to the of urban gardens and hedges capable of hosting animal species useful for the economy of ecosystems and of absorbing fine dust (5–10 ppm) harmful to health, to the redevelopment of sheds and terraces with hanging garden systems to contain heat losses. This network can also be connected by green, pedestrian and cycle paths, continuous and protected from vehicular traffic. The network idea also constitutes the reference model for local urban green policies. By virtue of what has been said, the need to abandon the sectoral approach that characterized rationalist urban planning to converge with conviction towards integrated and sustainable urban planning of networks is increasingly emerging.

2.2 Th Function of Greenery in the City

Increasing greenery by redesigning our cities allows us to have effective tools against the worrying climate change of recent years. Some measures, such as Agenda 21 and the Aalborg Charter, highlight its importance for improving the quality of the environment and life in cities. But what are the advantages of green? First of all it has a decorative function, useful for making cities beautiful, welcoming, livable, then ecological as it is decisive for improving the microclimate, for energy saving and for the sustainability of cities, and more. Over the years it has increasingly taken on a social function without underestimating the fact that it also increases citizens' safety and improves their mood, to the point of combating depression and loneliness. The time has come to measure ourselves with new parameters aimed at the quality of the place, the well-being of the community, a new governance. In Italian urban planning, the functions assigned to greenery have remained those prescribed by the urban planning standards referred to in Ministerial Decree 1444/1968 with the obligation of an abstract relationship between the quantity of areas to be allocated to services and those to be allocated to buildings for settlements, within the functional areas of the plan.. Law no. 10/2013 "Regulations for the development of urban spaces" did not produce obvious results. It is desirable to have a different urban planning culture that takes these spaces into account, with the extension of green roofs, right from the design phase, so as to contribute to improving both green and energy saving policies. We can and must underline the role of greenery from a bioclimatic point of view: the evapotranspiration produced by plants can contribute to a significant mitigation of summer temperatures in urban areas, without also neglecting the role of rivers. In line with the indications deriving from the planning, the objective of these reflections is to ensure the protection, safeguard and conservation of the river habitats and of the plant and animal species present and to guarantee - therefore - along the course of the rivers, the maintenance and /o the restoration of the ecological balances that characterize the habitats and which underlie their conservation. Achieving this conservation objective makes it necessary in particular to reconcile human activities that influence the status of species and habitats present in the river territory with their conservation. Precisely with a view to reorganizing the human activities present in the river territory to guarantee the protection of biodiversity, interventions are proposed aimed at promoting eco-compatible economic activities, correlated with the sustainable management of the natural environment and its resources, for the benefit of the economic development of the river territory. One of the main objectives is to support a use of the river territories modeled on the conservation needs of the territory itself to be considered as a priority. In the context of the European indications, in particular of Directive 2000/60, of the Birds and Habitats directives, in relation to the regulations in force regarding protection from and against waters, in the light of experiences gained in projects that provide for the protection and negotiated management of the landscape and the river environment, the proposal identifies the river system as the most suitable component for defining and developing the knowledge and dynamics of the "river world" in its territory, not only from an environmental point of view but also from a socio-economic point of view. co, promoting the governance of local development processes, involving the bodies responsible for this river redevelopment activity but above all the actors and subjects who are directly involved in this process starting from the municipalities involved and ending with private subjects. In particular, it is necessary to recognize rivers as entities with which we must coexist and perceive them as territorial, environmental, landscape and cultural references unifying the urban communities that find hospitality in its basin.

2.3 An Innovative Approach to Measure the Efficiency of Greenery. The Usefulness of Urban and Environmental Sustainability Indicators

It is not infrequently noted that some land use destinations contrast with the natural areas of the river course which favor different destinations that are more in keeping with the surrounding environment such as: areas of public interest, parks, leisure areas. ro and sports. The latter would allow us to obtain not only a status of environmental harmony, but also a better quality of life in the areas of the entire city. The objective of the proposed experimental model is to measure, through physical indicators (Fig. 1).

Fig. 1. Ruscello San Pietro a San Lorenzo Bellizzi (Calabria). Source: photograph taken by the author

geometric and vegetational, the performance of urban greenery (permeability, hedge lawns, shrubs, low - medium - tall trees) with respect to a series of functions (shading, permeability, production of oxygen O2, absorption of anhydride carbon CO2 and pollutants, acoustic insulation, creation of urban habitats and urban ecological micro-corridors to promote biodiversity, windbreak effect, division of spaces, visual isolation, etc. In addition to the well-known aesthetic and recreational functions, green areas contribute to mitigate pollution of the various environmental matrices (air, water, soil), improve the microclimate of cities and maintain biodiversity. However, these functions and benefits are poorly integrated into open space management policies. To make the urban sustainability and integrating it into territorial policies, it is useful to define a system of indicators that allows the level of sustainability of a city or metropolitan area to be assessed. The identification and use of sustainability indicators to support sustainable development policies are objectives widely recognized by the main international and European bodies. These indicators allow not only to draw up a diagnostic framework of the conditions present in a given context under examination, but also to monitor and verify the possible achievement of the pre-established objectives, thus constituting a valid cognitive and strategic tool for administrators, planners and citizens. Generally. Among the indicators developed at various levels by national and international bodies to pursue the objectives of urban sustainability, we recall urban green spaces, as well as that which refers to the accessibility of public green areas and local services, measured by the percentage of population living within 300 m of green areas of a size of at least 5000 m2 (parks, gardens, open spaces, equipment, usable private greenery…) and some basic services (healthcare, transport, education, food, greengrocers, etc.). Green areas are defined as: public parks, gardens or open spaces for the exclusive use of pedestrians and cyclists except green islands or traffic islands; equipment for outdoor sports, accessible free of charge to the public; private areas (agricultural areas, private parks) accessible to the public free of charge. The indicators that refer to urban greenery are: usable urban greenery, m2/inhabitant of usable greenery, excluding parks and protected areas and green areas, surface area of the various green areas on the total municipal surface area (m2/ha). Each requirement depends on a set of basic indicators; the requirements are measured using an index calculated as a weighted function of the basic indicators with respect to functionality (rare functions; children's and teenagers' play area; elderly space; dog enclosure; barrier-free paths), aesthetic quality (historical and artistic value, floristic variety; presence of water; context and background), safety (safe pedestrian and cycle access; fence), services and furnishings (toilets; drinking water; bicycle parking; benches and bins), pressure factors (traffic roads; railways; power lines; industries and warehouses) and maintenance (turf; paths; cleaning). Furthermore, the quality of these areas is essentially measured according to their usability. However, it is believed that the quality of green areas also strongly depends on the ecological and environmental dimension: phytosanitary state of plants, biodiversity, care and forestry interventions could be some of the basic indicators for measuring and evaluating the environmental quality of greenery, to be on which the social one depends. It is appropriate to remember that the importance of ecosystem services is very high as they influence and support human life and well-being with reference to the basic ones: breathing, drinking and eating, but also those, equally vital for quality of life, such as leisure, physical movement in the open air, psycho-physical balance. The theoretical model of "environmental services" is a very useful analysis tool because - by identifying in detail the various functions of nature and the multiple services it provides - it helps us to "translate" the benefits for society into concrete terms. The indicators mentioned above will be useful for evaluating sustainability and urban quality in order to improve the local context and environmental impact which is fundamental for supporting the well-being of the community. Natural based solutions can bring benefits in relation to the protection of species and habitats, as well as adaptation to climate change and disaster risk reduction through the implementation of urban gardens and allotments, green parks, pollinator sites, green corridors, wetland restoration, sustainable urban drainage systems or green walls and roofs. At the conclusion of this writing, it is appropriate to remember that on 21 December 2023, with decree 434, the national climate adaptation plan was approved, the main objective of which is to provide a national framework for the implementation of actions aimed at minimizing the risks deriving from climate change as possible, to improve the adaptive capacity of socio-economic and natural systems, as well as to take advantage of any opportunities that may arise with new climatic conditions.

Acknowledgements. This research work is the result of the research activities conducted within the following PNRR research project funded by the European Union – NextGenerationEU: "Pilot Project 4.6.1" "Climate adaptation plans for the reduction of the ecological footprint and ecological debt, aimed at improving the conservation and transformation response in terms of resilience and quality of life in the Calabrian and Lucanian urban and territorial systems" (Goal 4.6 - Planning for Climate Change to boost cultural and natural heritage: demand-oriented ecosystem services based on enabling ICT and AI technologies - Tech4You Innovation Ecosystem), Finanziato dall'Unione europea- Next Generation EU, Missione 4 Componente 2 CUP C33C22000290006 (ECS_00000009).

References

1. Giaimo, C., Salata, S.: Nuovi paradigmi per la pianificazione urbanistica: i servizi ecosistemici per il buon uso del suolo. Intervento presentato al convegno Un nuovo ciclo della pianificazione urbanistica tra tattica e strategia tenutosi a Urbanpromo - XIII Edizione Progetto Paese, Triennale di Milano, Milano nel 11 novembre, pp. 219–224 (2016)
2. Snyder, G.: The real work, interview e talks 1964–1979., edited by Wm Scott Mclean, A new directions book (1980)
3. Lynch, K.: Progettare la città. La qualità della forma urbana, Etaslibri, Milano, p. 273 (1990)
4. Gabrielli, B.: Introduzione. In: Lynch, K. Progettare la città. La qualità della forma urbana, Etaslibri, Milano, p. XI (1990)
5. Gasparrini, C.: L'attualità dell'urbanistica: Dal piano al progetto dal progetto al piano, Etaslibri, Milano, p.102 (1994)
6. Ratti, C.: La città di domani. Come le reti stanno cambiando il futuro urbano, Passaggi Einaudi, Torino, p. 9 (2017)
7. Catalano, G.: Il verde urbano., bene comune futuribile. In: AAVV *Cosenza*, a cura di Catalano Gilda e Sguglio Alfredo, Rubbettino, Soveria Mannelli Catanzaro, p. 7 (2022)

Urban Planning, Design, and Governance Environment for Urban Transformations Enhancing Ecosystem Services

Circular Economy, Bioeconomy and Ecosystem Services: The Craftsmanship Activities in Sicily

Vincenzo Provenzano(✉) and Ugo Lombardo

Department of Business Economics and Statistics DSEAS, The University of Palermo, Viale delle Scienze, 8, 90128 Palermo PA, Italy
{vincenzo.provenzano,ugo.lombardo}@unipa.it

Abstract. Ecosystem services linked to biodiversity conservation and sustainable natural resource use are underscored as a crucial research focus. Extensive literature describes the role of the circular economy as a protagonist in the challenges that will characterize global development in the coming years. The circular economy emerges as a key player in shaping the global growth trajectory, with a pivotal role in fostering sustainability within cities and societies. Its transformative impact is underscored by a shift from a linear economic model to a more adaptive transition model, emphasizing the importance of innovation in research and entrepreneurship. The paper subsequently looks into a pilot study conducted in Sicily, focusing intently on classifying economic activities within Sicilian craftsmanship. Its overarching objective is to discern those activities that resonate with the pivotal shift towards sustainability and a circular economy. The paper provides an overview of a broad study that integrates societal resilience, sustainable economic approaches, and ecosystem services. It focuses explicitly on Sicilian craftsmanship and its alignment with the transition towards a circular and bio-based economy.

Keywords: Ecosystem services · Bioeconomy · Craftsmanship

1 Introduction

Recognizing the importance of ecosystem services, this article proposes a pilot experiment using the Statistical Classification of Economic Activities in the European Community (NACE), Classification of Italian Economic Activities (ATECO), and Common International Classification of Ecosystem Services (CICES) to identify craftsmanship sector activities in Sicily aligned with the bioeconomy and those contributing to ecosystem services. This multidimensional approach offers a valuable framework for understanding the symbiotic relationship between circular economy practices [1, 2], the bioeconomy, and eco-system services in pursuing sustainable regional development. The structure of the paper is outlined: the first part defines the circular and bio-economy, exploring their connections and relevant sectors. The second part involves the analysis

"This article is supported by MUR, PRIN 2022 PNRR ECO-SET: A Multidisciplinary approach to plan ECOsystem SErvices for cities in Transition" – GRANT N. P202249BMJ_004.

F. Moraci et al. (Eds.): NMP 2024, LNNS 1272, pp. 93–106, 2025.
https://doi.org/10.1007/978-3-031-82927-7_6

of the NACE classification to identify bioeconomy activities and delves into ecosystem services and the CICES classification. The final part uses NACE classification and compares ATECO (ISTAT) and CICES classifications to analyze the distribution of Sicilian craftsmanship and identify, respectively, bioeconomy activities and ecosystem services impacting well-being. Additionally, GIS tools are mentioned as a means of creating maps to aid in the analysis and make it more transparent.

2 A Brief Literature Review of Circular Economy, Bioeconomy, and Ecosystem Services

The MacArthur Foundation's definition of the circular economy, centered on recycling and reuse through uninterrupted life cycles [3], sets the stage for its practical application. One convincing example is the conversion of food waste, turning what was once deemed waste into the foundation for new production cycles, effectively eliminating the concept of waste.

Contemporary literature advocates a paradigm shift towards more responsible and sustainable attitudes, emphasizing a systemic approach that considers the interplay between products, time, and regional context to achieve sustainability and minimize waste [4]. The circular economy, with its regenerative and restorative principles relying on renewable energy and creating new life cycles from waste, complements the bioeconomy.

The European Commission defines the bioeconomy as a comprehensive field covering sectors and systems reliant on biological resources, linking land and marine ecosystems, primary production, and economic and industrial sectors. This bioeconomy, closely associated with technological development and global economic growth, has evolved with concepts like the bio-based economy and circular economy [5].

The synergy between the bioeconomy and the circular economy is pivotal, leading to a "circular bioeconomy" [4–6]. This integrated approach emphasizes the interconnected progress of these concepts, highlighting the essential role of sustainably produced biomass in substituting non-renewable energy. European initiatives have already taken strides in promoting eco-innovative solutions to limit food waste production and encourage reuse, particularly in the Southern regions. Their impact extends beyond individual companies, creating a network of local businesses that contribute to well-being and economic returns in the area.

This networking effect represents both a challenge and an opportunity for regions seeking enhanced economic development by establishing green tech companies and promoting a sustainable, innovative, and circular regional context. Moreover, social enterprises and companies aligned with circular economy principles generate regional impacts and create new jobs and networks, acting as ecosystem services and providing direct and indirect benefits to the community and the territory [7].

3 NACE Classification for Bioeconomy and CICES Classification for Ecosystem Services

The European Commission's broad definition of the bioeconomy encompasses many activities, highlighting the importance of precisely delineating its constituent sectors. In this endeavor, the Statistical Classification of Economic Activities in the European Community (NACE) plays a crucial role, offering a comprehensive framework to identify and quantify the diverse economic activities associated with the bioeconomy [5].

The NACE classification reveals two primary categories within the bioeconomy dimension. Firstly, activities rooted in natural resources directly exploit biological resources, such as agriculture, fishing, and forestry, forming the primary sectors supplying biomass for subsequent processing. These activities fall under divisions A01 to A03, distinctly aligning with the fundamental pillars of the bioeconomy. Secondly, conventional manufacturing activities that further process biomass, including sectors like food processing, wood processing, textiles, and chemical products, represent the central part of bioeconomy sectors. The NACE classification includes these activities under Section C, encompassing divisions C10 to C22. This classification delineates traditional sectors utilizing bio-based inputs and emerging activities processing biomass as a substitute for fossil-based raw materials:

In addition, various service activities partially use processed biological resources. This list includes sectors in the following divisions:

a. D35 (electricity, gas, steam, and air conditioning supply);
b. F41 (construction), F42 (civil engineering);
c. G46 (wholesale trade), G47 (retail trade);
d. I55 (accommodation) and I56 (food and beverage service activities).

Simultaneously, the Common International Classification of Ecosystem Services (CICES), established in 2009 and now in its thoroughly revised version V5.1, provides an alternate classification system. Rooted in the concept that ecosystem services contribute to human well-being, CICES defines these contributions as 'what ecosystems do' for people, distinct from the goods and benefits derived from them. The cascade mode within CICES classifies final ecosystem services as the direct outputs of ecosystems—natural, semi-natural, or highly modified—directly impacting human well-being. This model considers the purposes or uses of people for different types of ecosystem services and the specific attributes or behaviors of ecosystems that support them (Fig. 1).

This model visualizes CICES's overarching goal—to categorize final ecosystem services as the vital contributions of ecosystems (living systems) to human well-being.

Refining the Common International Classification of Ecosystem Services (CICES) structure, initially proposed and further developed in version V5.1, prioritizes internal consistency. This hierarchical structure acknowledges the diverse thematic and spatial scales at which individuals operate, facilitating flexibility in aggregating classes based on user preferences. The compositional structure of CICES delineates the hierarchy into "Sections," encompassing three main categories: "provisioning," "regulation and maintenance," and "cultural".

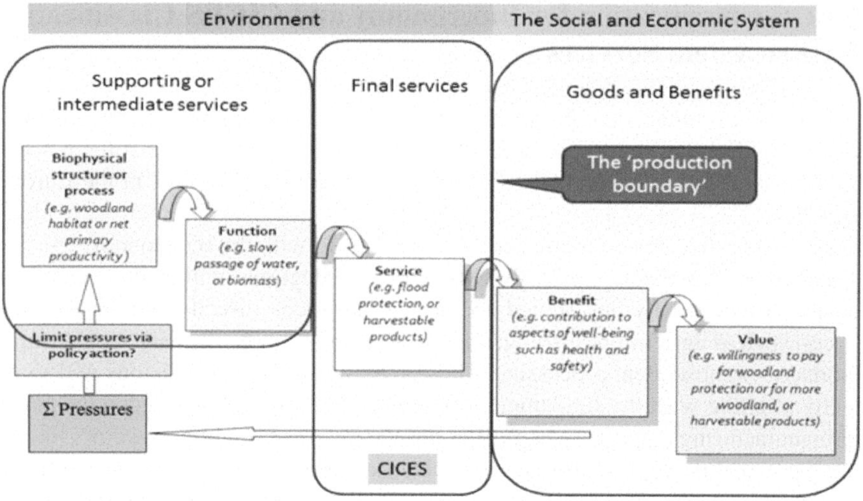

Fig. 1. The cascade model (Potschin and Haines-Young, 2016b). Note: The cascade model is derived from the Common International Classification of Ecosystem Services (CICES) V5.1 Guidance on applying the Revised Structure.

It is possible to aggregate classes differently through this hierarchy, allowing users to select the most beneficial [8]. The composition of the CICES classification is shown in the following figure (Fig. 2).

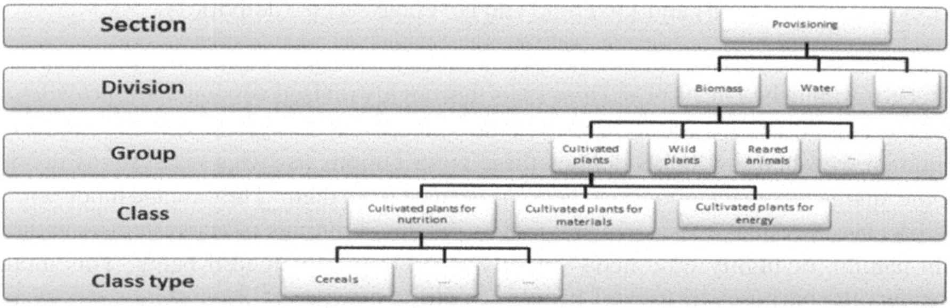

Fig. 2. The Structure of CICES Classification V5.1. Note: The Structure of CICES is derived from the Common International Classification of Ecosystem Services (CICES) V5.1 Guidance on applying the Revised Structure.

At the broadest level, the "Sections" provide a foundational framework, further subdivided into "Divisions," "Groups," and "Classes." The figure exemplifies the hierarchical structure focusing on Provisioning Services. As users progress from Section to Division, Group, and Class, services become increasingly detailed while remaining nested within overarching categories [7].

This hierarchical design aims to cater to users' diverse needs, allowing them to select the appropriate level of detail for their applications [9]. The higher categories in

CICES are exhaustive in rendering the classification applicable, covering all elements recognized as ecosystem services by people [10].

The structure also incorporates a sense of "taxonomy," indicating conceptual similarity among elements within the same Group or Class, enhancing understanding of how these services are utilized. CICES, being a classification system rather than an arbitrary nomenclature, aligns with the requirements for this pilot study in Sicily. The upcoming sections will reveal the data and results of applying this refined structure.

4 Methodology and First Results

The decision to focus on enterprises in the craftsmanship sector in Sicily stems from the sector's dynamic growth, as evidenced by an increase in active enterprises and their widespread distribution across the region. According to data from the Chamber of Commerce of the Marche region [11], artisanal enterprises in Sicily grew from 71,074 in December 2019 to 72,022 in December 2022, indicating sustained growth despite the challenges posed by the pandemic. The Ente Bilaterale Artigianato Siciliano (Bilateral Craftsmanship Sicilian Entity, EBAS) saw a notable increase in artisan enterprises from 3,427 in 2019 to around 10,016 in 2022 [12]. This growth positions the craftsmanship sector as a pivotal reference point for exploring the transition towards a bioeconomic model and identifying ecosystem services that contribute to well-being in the region. Based on a sample of 10,016 craftsmanship enterprises in Sicily, the analysis provides a comprehensive overview derived from EBAS data, representing approximately 14% of all active enterprises in Sicily based on Chamber of Commerce data. The provincial distribution of the reference sample highlights a concentration of companies in Palermo, Catania, Trapani, and Messina, as shown in Fig. 3.

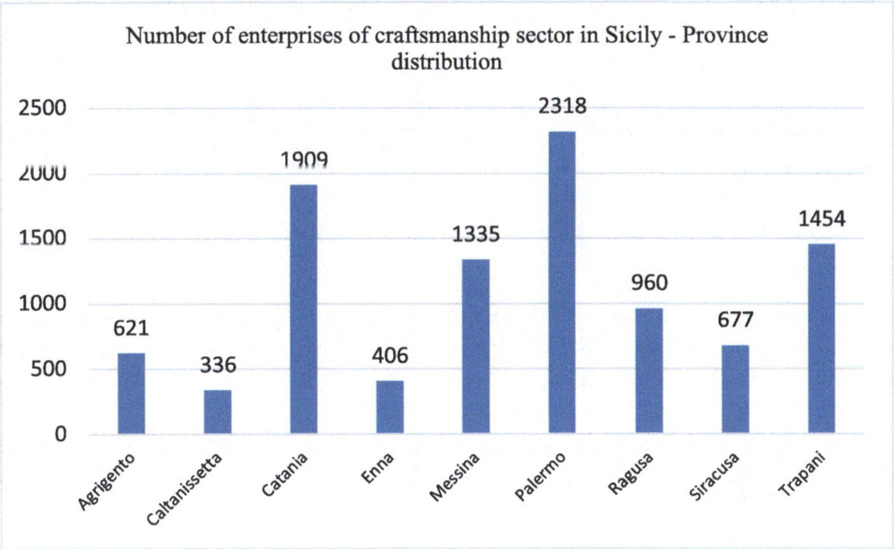

Fig. 3. Authors' elaboration based on EBAS data 2023.

Despite its smaller territorial size, Ragusa boasts more enterprises (677) than some larger provinces, emphasizing its significance in diverse areas. Before investigating the results, a brief methodological explanation is warranted. ISTAT and ATECO [13] codes were cross-referenced to identify bioeconomic companies, aligning them with NACE classifications. The CICES classification was employed to analyze ecosystem services, considering its highest hierarchical level—Cultural, Provisioning, and Regulation and Maintenance sections.

4.1 Sicilian Craftsmanship Enterprises in Bioeconomy

According to the NACE classification, 2,529 companies in the sample, approximately 25%, exhibit activities aligned with the bioeconomy. These companies demonstrate a clear inclination towards sectors that promote reuse, waste reduction, and innovation in bio-based materials, designs, and business models. The provincial distribution of bioeconomic enterprises shows how Palermo, Catania, and Trapani lead the way, as shown in Fig. 4.

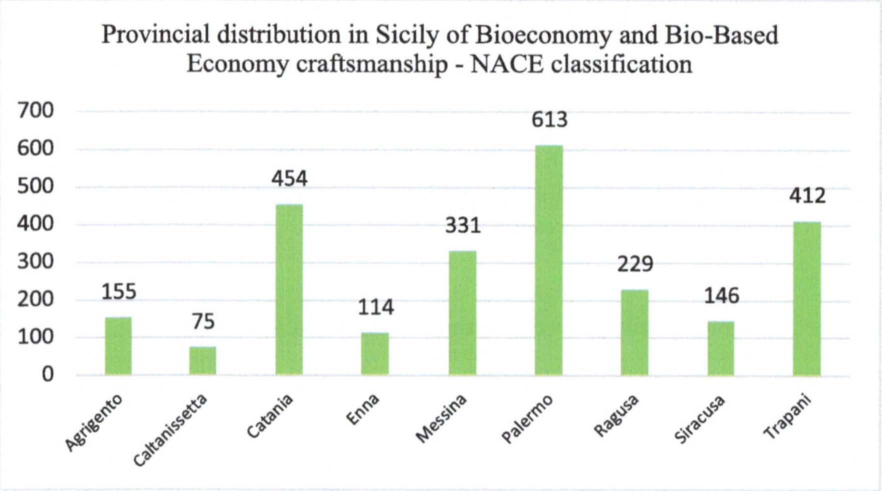

Fig. 4. Authors' elaboration on EBAS data and NACE classification 2023.

The subsequent graph (Fig. 5) details the distribution of companies across bioeconomic activities according to the NACE classification. Notably, 54% of companies fall under Division C10 (food products), 18% under Division I56 (food and beverage service activities), and 8% under Division C16 (wood and wood products).

These results underscore the sector's bioeconomic potential and provide valuable insights for policymakers and stakeholders seeking to foster sustainable economic development in Sicily.

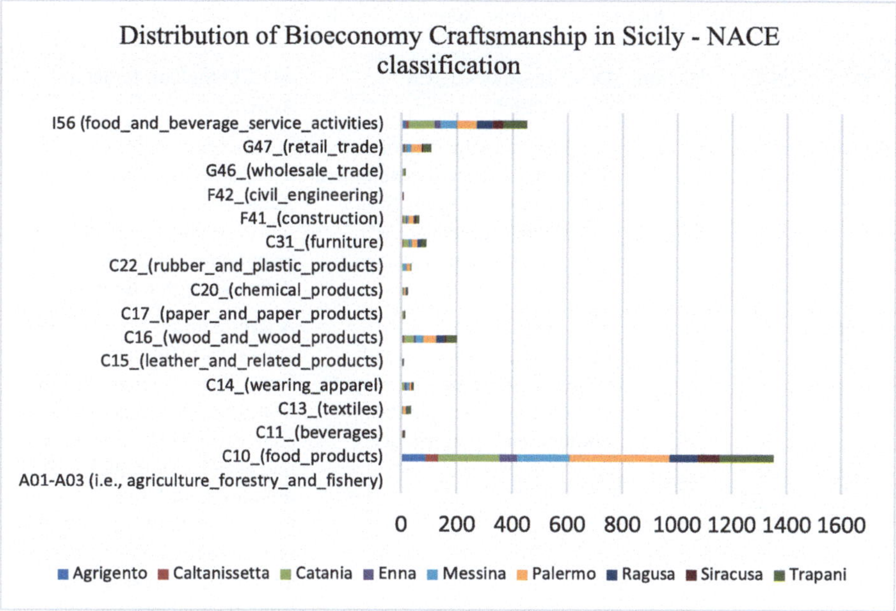

Fig. 5. Authors' elaboration on EBAS data and NACE classification 2023.

4.2 The Presence of Ecosystem Services in Sicily

Below is a table summarizing the Sicilian economic activities falling within the definition of the three ecosystems for each appropriately described section. In the case of the Cultural ecosystem, enterprises with ATECO codes I, R, and S are included. This section embraces all non-material, typically non-rival, and non-consuming products of ecosystems (both biotic and abiotic) that influence the physical and mental states of individuals. Activities with codes A, C, G, I, and N are included in the Provisioning ecosystem. This section encompasses all activities and sectors that generate nutritional and non-nutritional material, energy products from living systems, and abiotic products (including water). Finally, the Regulation and Maintenance section is the broadest and may include sectors of the other ecosystems. Indeed, it has economic activities with codes B, C, E, F, G, H, I, J, K, L, N, Q, and S. The reason why so many sectors are included is that this section encompasses all ways in which living organisms can mediate or moderate the environmental surroundings that affect human health, safety, or comfort, along with their abiotic equivalents.

By describing each activity through the ISTAT tax code, it was possible to verify whether it falls within the definition of this ecosystem (Table 1).

Moving on to the distribution of businesses by ecosystem at the provincial level, we can observe from the following graph (Fig. 6) that approximately 59% of enterprises fall into the Regulation and Maintenance section. In comparison, 21% of enterprises fall into the Cultural section, and 20% into the Provisioning section. Once again, Palermo, Trapani, and Catania are the provinces with the highest number of businesses for each ecosystem. The most stimulating aspect is also regarding Ragusa, which, despite being

Table 1. Sicilian Economic Activities in CICES Classification

Section CICES Classification	Description of CICES classification	ATECO Sicilian Economic Activities
Cultural	All the non-material, normally non-rival, and non-consumptive outputs of ecosystems (biotic and abiotic) affect people's physical and mental states	I (Information and communication services); P (Education); R (Artistic, sporting, entertainment, and entertainment activities); S (Other service activities)
Provisioning	This Section covers all nutritional, non-nutritional material, and energetic outputs from living systems and abiotic outputs (including water)	A (Agriculture, Forestry, and Fishing); B (Extraction of Minerals from Quarries and Mines); C (Manufacturing Activities); G (Wholesale and retail trade; repair and motorcycles); I (Accommodation and catering services activities); N (Rental, Travel agencies, Business support services)
Regulation and Maintenance	All how living organisms can mediate or moderate the ambient environment that affects human health, safety, or comfort, together with abiotic equivalents	B (Extraction of Minerals from Quarries and Mines); C (Manufacturing Activities); E (Water supply; sewerage networks, waste management, and sanitation activities); F (construction); G (Wholesale and retail trade; repair and motorcycles); H (Transport and warehousing); I (Accommodation and catering services activities); J (Information and communication services); K (Financial and insurance activities); L (Real estate activities); Professional, scientific and technical activities); N (Rental, Travel agencies, Business support services); Q (Health and Social Care); S (Other service activities)

smaller than other provinces, has several companies per ecosystem that is only lower than the fourth province, Messina. Agrigento and Caltanissetta, on the other hand, are the last in terms of the number of businesses per ecosystem.

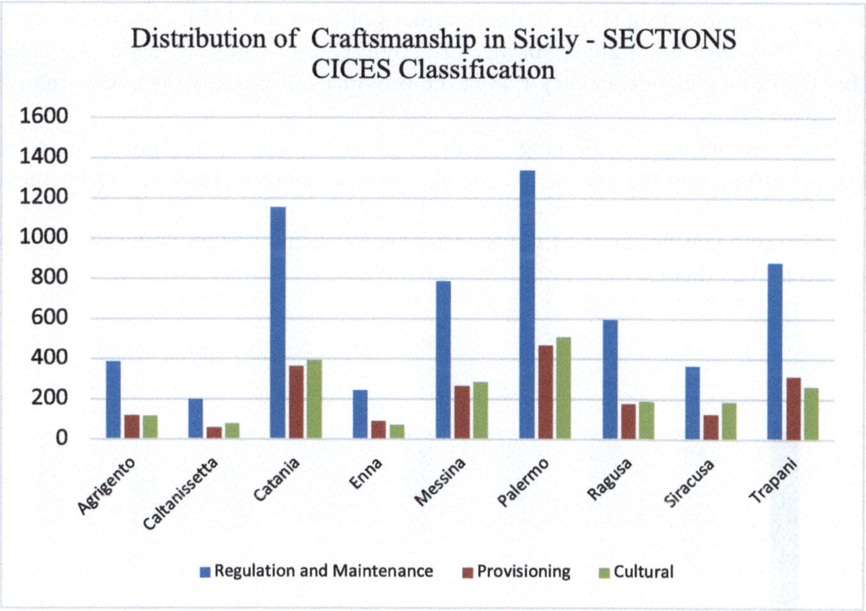

Fig. 6. Authors' elaboration based on EBAS data and NACE classification 2023.

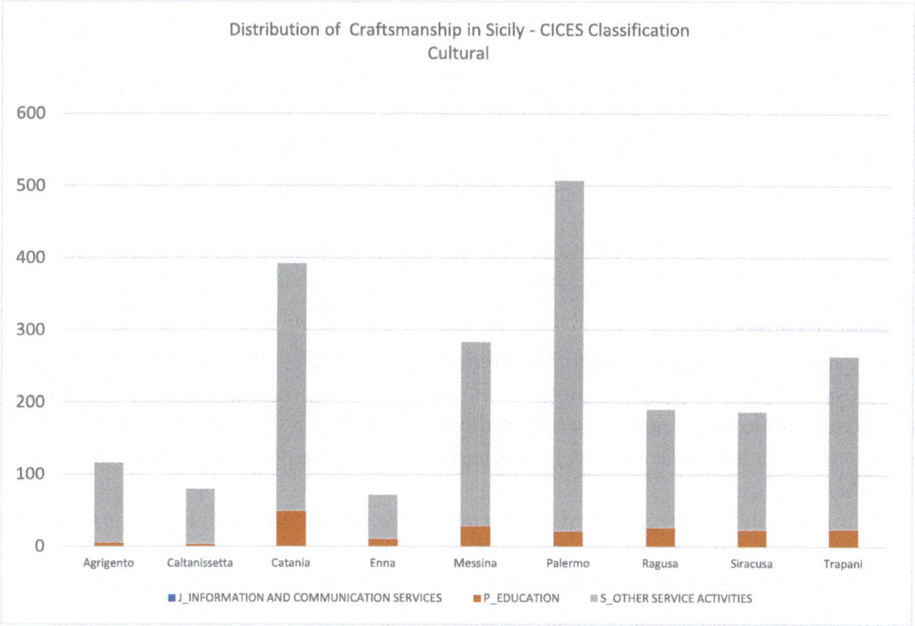

Fig. 7. Authors' elaboration based on EBAS data and NACE classification 2023.

In the Cultural section (Fig. 7), the provinces of Palermo (485), Catania (342), and Messina (254) have the highest number of activities in Category S, while the highest number of businesses in Category P is in the provinces of Catania (49), Messina (29), and Ragusa (27).

In the Provisioning section (Fig. 8), the highest number of enterprises are related to manufacturing, and the provinces with the most remarkable presence of businesses defining this ecosystem are Palermo (366), Catania (234), and Trapani (207). What stands out is the significant presence of the S sector in Agrigento, which is the top province compared to the others.

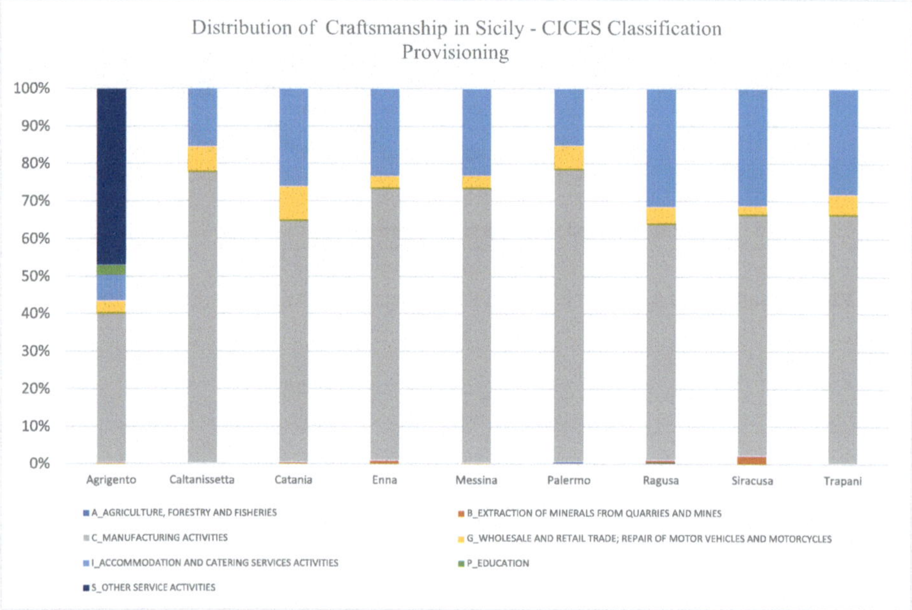

Fig. 8. Authors' elaboration based on EBAS data and NACE classification 2023.

Regarding the Regulation and Maintenance section (Fig. 9), the main activities in this ecosystem are related to sectors C, G, and F. Therefore, the provinces with these enterprises and a significant presence in this ecosystem are Palermo, Catania, Trapani, and Messina. Ragusa follows closely, while Agrigento is the only one without businesses in sector F.

Ultimately, the study generated and juxtaposed several maps (Fig. 10) to garner a spatial perspective on the distribution of ecosystems and ecosystem services across Sicily. Additionally, Geographic Information System (GIS) technology facilitated a comparative analysis between a map illustrating the concentration of all Sicilian craftsmanship enterprises in 2022 and maps representing the presence of the ecosystems. Kernel Density Estimation was used to estimate the concentration, and the radius used was the surface area of Sicily implemented with the Raster and Map Units mode to standardize the concentration relative to the overall size of Sicily.

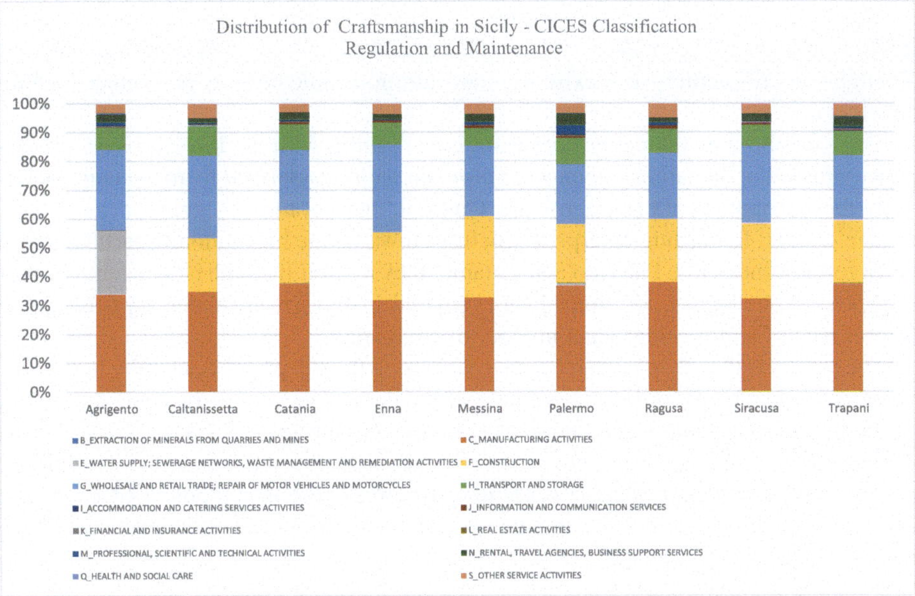

Fig. 9. Authors' elaboration based on EBAS data and NACE classification 2023.

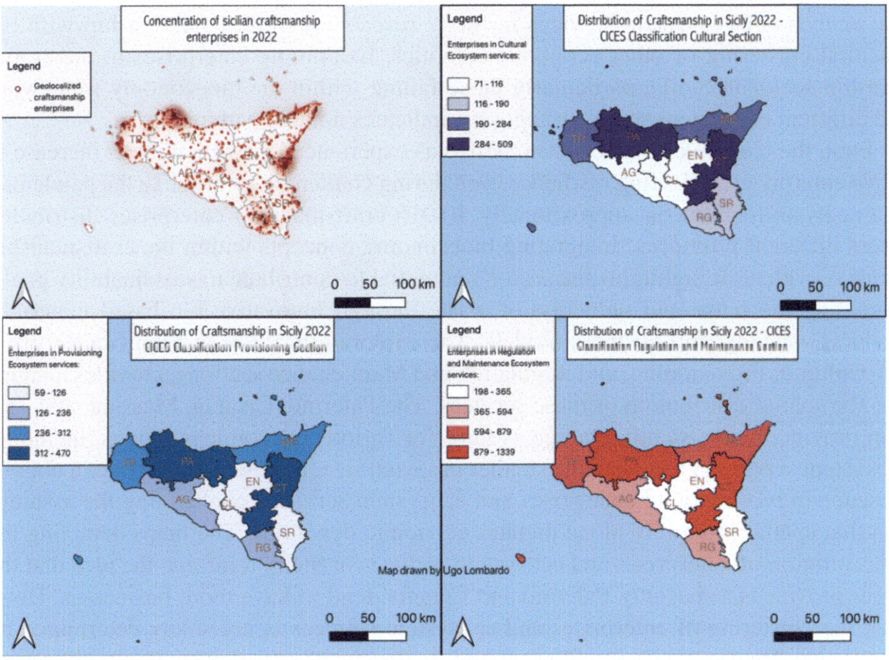

Fig. 10. Author's elaboration based on EBAS data and NACE classification 2022.

The bulk of business concentration is primarily observed in Palermo, Catania, Messina, and Trapani provinces. Ragusa also exhibits a notable concentration, whereas although numerous enterprises exist, the concentration could be more pronounced within the territory.

This corroborates the pattern of businesses aligning with the three ecosystems primarily situated in the larger provinces. This observation is derived from the maps delineating the distribution of firms across the Cultural, Provisioning, Regulation, and Maintenance sectors. Regarding enterprises inclined towards the bioeconomy, particularly in the larger provinces like Palermo and Catania, they exhibit the highest concentration of businesses. Nonetheless, additional factors beyond spatial size are necessary to elucidate the presence of a more significant number of companies.

The reason is that Trapani and Ragusa are among the top provinces for the number of enterprises considered for bioeconomics and the three ecosystems. Instead, Agrigento, similar in size to Catania and Palermo, is consistently among the last provinces in terms of concentration and number of businesses. This could suggest that the spatial proximity to larger provinces with positive economic dynamics (the development of economic activities and the presence of companies) generates a positive and expansive effect, influencing the surrounding territory.

5 Conclusions

The analysis of ecosystem services in Sicily reveals a complex relationship with the potential clustering of other economic activities. Examining enterprises in the craftsmanship sector in Sicily, particularly those falling within the bioeconomy framework and different ecosystem service categories, indicates notable patterns.

First, the craftsmanship sector in Sicily has experienced growth, with an increase in active enterprises, reflecting resilience even during challenging times like the pandemic. The analysis focused on approximately 10,016 craftsmanship enterprises distributed across different provinces. Integrating bioeconomy concepts within the craftsmanship sector is evident. It highlights the sector's potential to contribute to sustainability goals, promoting the reuse and reduction of waste through innovative bio-based materials, products, designs, and business models. The exploration of CE in Sicily, categorized into Cultural, Provisioning, and Regulation and Maintenance sections, provides insights into the spatial distribution of these services. The Palermo, Catania, Messina, and Trapani provinces consistently emerge as hubs for various economic activities, including ecosystem services. Despite being smaller in spatial size, Ragusa demonstrates a notable presence in bioeconomic enterprises and ecosystem services, challenging the assumption that spatial proximity alone dictates economic dynamics. The maps depicting the concentration of enterprises and ecosystem services in Sicily reinforce the idea that the larger provinces, especially Palermo and Catania, tend to have more businesses. However, the clustering of enterprises and ecosystem services is not solely determined by spatial size. Despite their size, Trapani and Ragusa stand out as provinces with significant economic activity, suggesting that positive economic dynamics in larger provinces might positively influence nearby territories. The relationship between ecosystem services and economic activities in Sicily underscores the importance of understanding the interplay between different sectors. Spatial dynamics, economic clustering, and the integration of sustainable practices, such as those seen in the bioeconomy and craftsmanship sector, contribute to a nuanced perspective on regional development [14] and the potential for creating well-being and economic progress [15]. The analysis suggests that fostering sustainable practices and innovation in various economic sectors can have positive spillover effects, creating a more resilient and interconnected regional context.

Acknowledgments. This research work is the result of the research activities conducted within the following PNRR Research Project funded by the European Union – NextGenerationEU: "ECO-SET - A Multidisciplinary approach to plan Ecosystem Services for cities in Transition" research project, Finanziato dall'Unione europea- Next Generation EU, Missione 4 Componente 2 - CUP B53D23029450001.

References

1. Gardetti, M.A.: Introduction and the concept of circular economy. In: Circular Economy in Textiles and Apparel, pp. 1–11. Woodhead Publishing (2019)
2. Gura, K.S., Nica, E., Kliestik, T., Puime-Guillén, F.: Circular economy in territorial planning strategy: incorporation in cluster activities and economic zones. Environ. Technol. Innov. **32**, 103357 (2023)
3. Ellen MacArthur Foundation, Towards the circular economy Vol. 1: an economic and business rationale for an accelerated transition (2013)
4. Scaffidi, F.: Regional implications of the circular economy and food greentech companies. Sustainability **14**(15), 9004 (2022)
5. Kardung, M., et al.: Development of the circular bioeconomy: drivers and indicators. Sustainability **13**(1), 413 (2021)
6. Tan, E.C., Lamers, P.: Circular bioeconomy concepts—a perspective. Frontiers in Sustainability **2** (2021)
7. Haines-Young, R., Potschin, M.B.: Common International Classification of Ecosystem Services (CICES) V5.1 and Guidance on the Application of the Revised Structure (2018). www.cices.eu
8. Aryal, K., Maraseni, T., Apan, A.: Spatial dynamics of biophysical trade-offs and synergies among ecosystem services in the Himalayas. Ecosyst. Serv. **59**, 101503 (2023)
9. Tormáné Kovács, E., Centeri, C.: Assessment of ecosystem services at different scales. Sustainability **15**(24) (2023)
10. Czúcz, B., et al.: Where concepts meet the real world: a systematic review of ecosystem service indicators and their classification using CICES. Ecosystem Services **29**, 145–157 (2018)
11. Camera di Commercio delle Marche, Open Data explorer Demografia delle Imprese in Italia. https://opendata.marche.camcom.it/stock-imprese.htm?url= https://opendata.marche.camcom.it/data/Stock-Imprese-Artigiane-Attive-Italia.json&f2v=22&color=F26522
12. Provenzano, V., EBAS in Sicilia – Inizi, evoluzione e tendenze 1993–2023, Report, Arti Grafiche Etnea Molino, Catania, October (2023);
13. Istituto Nazionale di Statistica ISTAT. https://www.istat.it/it/archivio/17888#codesearch

14. Griffith, D.A.: Spatial Autocorrelation and Spatial Filtering: Gaining Understanding through Theory and Scientific Visualization. Springer, Cham (2003). https://doi.org/10.1007/978-3-540-24806-4
15. Deng, X., Wang, Y., Song, M.: Development geography for exploring solutions to promote regional development. Geography and Sustainability **4**(1), 49–57 (2023)

A Reformulation of the Planning Process in the Light of Ecosystem Services: The European *Blue and Green City* Project

Alessandra Barresi(✉)

Mediterranean University of Reggio Calabria, Reggio Calabria, Italy
allessandra.barresi@unirc.it

Abstract. The desire to attribute to urban planning the ability to interface with the needs and problems of the city and the contemporary territory has meant that in recent years a large part of the scientific literature and urban planning practice has been directed towards the identification of actions that can contribute to resolving the climate issue. Having scientifically ascertained the importance of biodiversity in the process of climate mitigation and adaptation, it has become urgent and indispensable to start a process of safeguarding, valorising and restoring natural ecosystems, both in urban and territorial contexts, identifying the services they produce as the main drivers of biodiversity. The transformation processes determined in the urban sphere by the urban plan can increase the supply of ecosystem services as well as reduce their diffusion in the territory; the plan, even with possible compensation actions, must instead guarantee a high supply of ecosystem services. It is in this context that this contribution is made, which emphasises the role of urban planning in the safeguarding of biodiversity, in particular by referring to the *Blue and Green City* project, a good example of urban planning that, through the valorisation of Green and Blue Infrastructure, initiates the regeneration of the social system, landscape and environmental system of a sample area in the Piedmont Region through a trans-scalar approach that, by crossing the Vulnerability and Resilience factors with the abundance or lack of Ecosystem Services, identifies the most suitable planning actions for the valorisation of the infrastructures themselves as the main providers of ecosystem services at the various scales of the territorial and urban space.

Keywords: Ecosystem Services · Blue and Green Infrastructure · Urban and Territorial Planning

1 Foreword

The research that is being developed, in which this contribution is included, is grounded in the worldwide focus on the importance of safeguarding biodiversity for the survival of our planet. The interest in biodiversity attracts the attention of various profiles, from the institutional to the scientific and research ones, since the awareness has matured that only by guaranteeing the maintenance of an adequate level of biodiversity is it

F. Moraci et al. (Eds.): NMP 2024, LNNS 1272, pp. 107–127, 2025.
https://doi.org/10.1007/978-3-031-82927-7_7

possible to succeed in mitigating environmental risks of various kinds (hydrological, geological, climatic) that undermine the security of life on the entire planet, both in the most developed and in the most backward areas. Together with the loss of biodiversity, climate change, the modification of the biogeochemical cycles of water, atmosphere and soil, and the spread of pandemics show that there is a limit to the Earth's capacity to withstand the degradation of ecosystems (Rawtorth, 2012). Increasing natural capital and reactivating ecosystem cycles and processes is crucial to meeting the challenges of ongoing change. Natural resources and ecological functions expressed by natural capital provide free ecosystem services (ES) to human society that underpin the development of landscapes where human societies thrive and their activities take place (Gibelli et al., 2022).

Protecting, valuing, restoring biodiversity and keeping ecosystems healthy is essential for the essential services and benefits they can provide. Through the flow of ecosystem services, nature can offer long-term, smart, cost-effective and integrated solutions to numerous challenges, such as climate change mitigation and adaptation, disaster risk management and pollution, with positive impacts on the health and well-being of individuals and communities. Ecosystem services are the pillars on which the resilience of landscape systems is built, so territories need flexible tools built on diversity, resources, inherent characteristics and capacities to adapt to change. The rational and responsible use of land resources, together with decisions on land use, can help to build liveable spaces, to mitigate the effects of ongoing climate change, to find a balance between consumption, renewal and reuse of resources functional to a healthy environment. In this perspective described above, it is possible to think of a new form of urban planning that will be substantially aimed at the transformation of the existing environment through a process that is strongly contextualised and particularly attentive to the natural components of the urban space that, in order to revitalise the natural capital, need to be strongly connected through the creation of a true network, now also recognised at a European level, as green and/or blue infrastructure. This becomes the focus of the proposed research, i.e. rediscovering the link between the plan and the enhancement of ecosystems with a consequent increase in the provision of the services they produce to increase human wellbeing.

The issue of the protection and restoration of ecosystem services is an important novelty in territorial and urban planning, especially in the perspective of regeneration; assessing the performance of ecosystem services becomes fundamental for developing urban and territorial regeneration scenarios and for measuring the sustainability effects of choices (Santolini, 2022). Ecosystem services are one of the new interpretative paradigms that are supporting the transition of the urban planning discipline towards greater environmental values. The potential of the valuation of ecosystem services is precisely that of estimating, on the basis of use transformations, the ecosystem value differentials of transformed soils in both quantitative and qualitative terms. The interest on the valuation of ecosystem services in the institutional and academic debate, therefore, is high precisely at a time when one of the criticisms levelled at the theoretical advancement and practical development of ecosystem valuation is that of not being able to support the effective construction of the spatial and regulatory structure of the urban plan in response to territorial vulnerabilities (Caldarice, Salata, 2019).

Today, there is an increasing awareness of the weight that urban areas assume in altering environmental balances, and ecological issues are now becoming inescapable priorities for territories in transition from vulnerability and, therefore, ideal environments for ecological research. An important contribution to this renewed interpretative dimension of the discipline is provided by Ecosystem Services. Since the 1990s, Ecosystem Services have gained a growing consensus both with regard to the importance of their assessment in supporting natural resource management decisions in the context of land-use planning and their role in counteracting the loss of biodiversity caused largely by increasing territorial vulnerabilities. In this perspective, ES constitute a support for urban and spatial planning and design practices as they enable the understanding and evaluation of the environmental effects resulting from land use transformations and their consequent economic and social impacts. The assessment of ES implies, with respect to land use transformations at different time thresholds, the ecosystem value differentials in both quantitative and qualitative terms. Unfortunately, the connections between the theoretical knowledge of ES and their concrete use in urban planning and design are still weak (Caldarice, Salata, 2019). From this awareness derives the desire to know, understand and analyse those urban planning instruments and studies that in various ways have made use of the concrete use of ES in their structuring; several of them have been taken into consideration and of varying nature that have affected urban and regional levels.

The current Italian panorama presents, in fact, several experiences of integrating ecosystem services in urban and territorial planning processes at different scales, which are based on the increasingly felt need to have an integrated and multifunctional approach to land management, aimed at reducing the overall vulnerability of the system and the loss of ecological functionality and the services and benefits that ecosystems can provide (Santolini and Morri, 2017). These experiences have experimented with different issues including that of identifying and developing innovative schemes and tools and methodologies for the integration of ecosystem services in the assessment practices and planning and design of the issue of protection and restoration of ecosystem services (D'Ambrogi, Gibelli 2022).

In particular, in this contribution, we focused on the Blue Green City project, funded by the Interreg Europe 2014_2020 programme, which has the overall objective of improving policies that promote the value of green and blue infrastructure as an integral part of a local or regional natural heritage conservation strategy, and which after an initial period (of approximately two and a half years) during which each partner's policy instruments were assessed, potential barriers and drivers for the uptake of green and blue infrastructure, and exchanged practices and experiences to develop and improve policies to support the construction of these infrastructures in the participating cities and regions, Guidelines were drawn up, that build on the outcomes of the previous European project LOS_DAMA!, with which they share the methodologies and study area and capitalise on the results of the governance activated with the Corona Verde strategic project for the realisation of a metropolitan green infrastructure. The result of these Guidelines is a comprehensive guideline for the strategic planning, design and sustainable maintenance of green and blue infrastructure with an integrated, participatory and place-based approach, addressed to local public administrations, technicians and stakeholders.

The process of deepening the knowledge of these experiences of plans is aimed at achieving the first milestone, the construction of a catalogue of ES, of Goal 4.6 “Climate change planning to promote cultural and natural heritage: Demand-driven ecosystem services based on ICT and AI enabling technologies” of the Pilot Project “Climate adaptation plans for the reduction of the ecological footprint and ecological debt aimed at improving the conservation” within SPOKE 4 “Resilience and accessibility in the enhancement of local cultural and natural heritage” of the TECH4YOU Research Project “Technologies for climate change adaptation and quality of life improvement”. In addition to this introductory part, in which the subject under study is contextualised and specified, the paper is divided into a second paragraph in which the concept of ecosystem services is analysed by referring to the existing scientific documentation, a third paragraph in which a critical-interpretative summary of the Green Blue City project and the consequent Guidelines is proposed, and a last paragraph containing some concluding reflections. Finally, it should be emphasised that the first milestone of the Pilot Project, i.e. the construction of a data catalogue for the systematisation of ecosystem services over two selected areas in the Basilicata and Calabria regions, takes place according to Action 5 of the European Biodiversity Strategy. Similar to Natural Capital, the EU prioritises biodiversity, the loss of which can affect the resilience of ecosystems by compromising the provision of ecosystem services. The new strategy is one of the key pillars of the European Green Deal, containing an Action Plan that sets the goal of establishing protected areas for at least 30 per cent of seas and land, restoring degraded ecosystems through sustainable agriculture, halting the decline of pollinators, restoring rivers to a free-flowing state, reducing pesticides by 50 per cent and planting trees.

2 Evolution of the Meaning and Classification of ES at Territorial and Urban Scales

2.1 Ecosystem Services at the Territorial Scale (PTCP Lecco, 2022)

The concept of ecosystem services took off, in parallel with Urban Ecology (a discipline for the study of urban ecosystems), in the 1980s, and the first publication containing this term, as indicated by most studies on the subject, is entitled “Extinction: the Causes and Consequences of the Disappearance of Species” (Ehrlich P.R, Ehrlich A.H., 1981). The first classifications are those made by Robert Costanza; in his article “The value of the world’s ecosystem services and natural capital”, Ecosystem Services are represented by the goods, materials, energies and information of natural capital that, combined with the artifacts and services of human capital, produce human well-being (Costanza, 1997) (Fig. 1).

By biologist G. Daily, ecosystem services are defined as the conditions and processes through which natural ecosystems and the species that create and sustain them make human life possible. Complementing this definition is the clarification that ecosystem services, in addition to producing goods, constitute the current primary life-supporting functions, such as sanitation, recycling and regeneration, also encompassing many intangible aesthetic benefits and cultural benefits. The services listed are as follows: air and

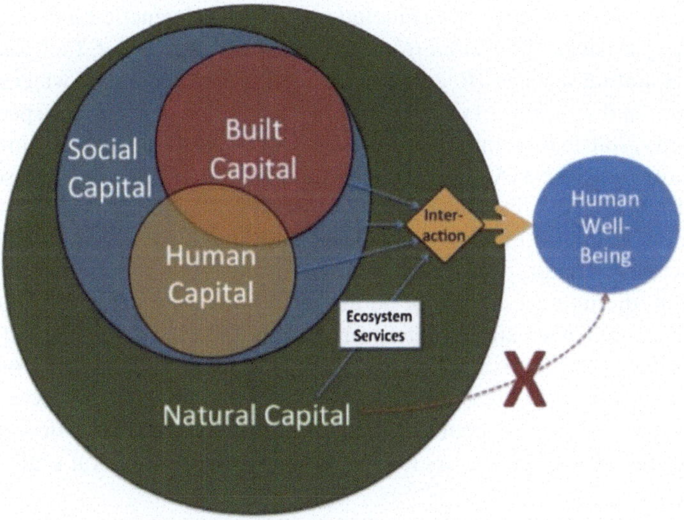

Fig. 1. A schematic representation of the aforementioned concept that renders the relationship between ecosystem services and different capitals, taken from a more recent article by R. Costanza et al. (2014), is given (PTCP Province of Lecco) (Source: PTCP Province of Lecco - Monograph G - Ecosystem Services, https://www.provincia.lecco.it/prlecco-media/2022/08/03_Monograph_G_Services_Ecosystem_2022.pdf.).

water purification; flood and drought mitigation; toxicity reduction and waste decomposition; soil and fertile soil generation and replenishment; pollination of crops and natural vegetation; control of most potential pests for agriculture; seed dispersal and nutrient processing; preservation of biodiversity; protection from exposure to the sun's ultraviolet rays; partial stabilisation of climate; mitigation of temperature extremes and wind and wave force; support for diverse cultures; and provision of aesthetic and intellectual stimuli that elevate the human spirit. (Daily, 1997).

In 2005, the Millennium Ecosystem Assessment project was developed for the United Nations under the Environment Program (UNEP) to analyse how ecosystems have evolved in relation to human activities and the resulting impacts, in terms of welfare conditions, and then to identify intervention strategies for sustainable development (Ma, 2005). The study, coordinated by R.T. Watson and A.H. Zakri, includes a survey of ecosystem service categories.

The MA proposes a very simple definition of eco-systemic services as the benefits that people obtain from ecosystems. This approach, which puts the human population at the centre, derives from the project's aims of assessing the effects on ecosystem balances that have occurred in recent times and the resulting impact on socio-economic health and well-being so as to incentivise ecosystem conservation policies.

The classification of eco-systemic services is based on a distinction of them according to the type of benefit and is divided into the following four categories: support; supply; regulatory; and cultural (Fig. 2).

Supporting services (*Supporting*) include habitats and genetic biodiversity; supply or provisioning services (*Provisioning*) provide goods (food, water, fuel, and other raw materials); regulating services (*Regulating*) deal with climate, air and water quality, and soil formation; and cultural services (*Cultural*) refer to non-material aspects, such as cultural identity, spirituality, intellect, aesthetic values, and recreational dimensions.

At the same time, the MA defines the relationship between ecosystem services and societal well-being, according to a scheme in which flows from the former to the latter are highlighted, based on the general concept that human well-being depends on the services provided by nature.

In addition, the interaction cannot be traced only to the natural and anthropogenic dimensions because some external determinants condition the dynamics and change the relationships; these are, for example, changes in land uses and land cover or the effects of climate change. For this reason, the management of ecosystem services and their enhancement must take into account future scenarios and the aforementioned dynamics, in the spatial dimension (global, regional, local) and at the temporal scale (short- and long-term).

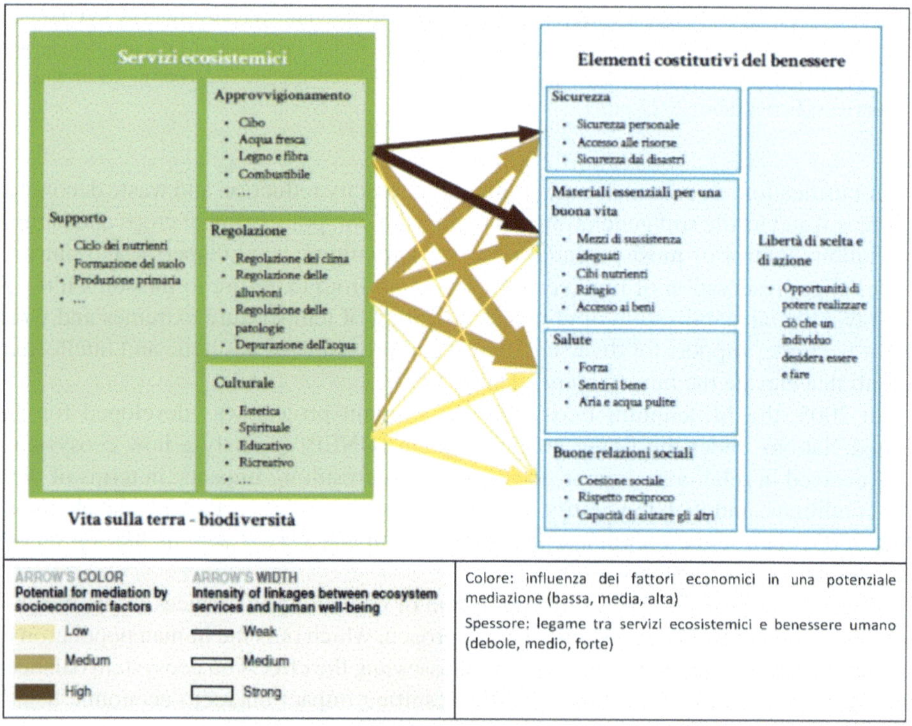

Fig. 2. Relational diagram between ecosystem services and the human well-being dimension. (PTCP Province of Lecco) (Source: PTCP Province of Lecco - Monograph G - Ecosystem Services, https://www.provincia.lecco.it/prlecco-media/2022/08/03_Monograph_G_Services_Ecosystem_2022.pdf.)

Following the definition of ecosystem services elaborated by the Millennium Assessment there are different views in subsequent years that differ from the proposed classification in that it does not distinguish between the means used to produce the goods and services and the goods themselves.

Among the most interesting definitions, subsequent to that of MA, is the one proposed by Fisher et al. who, in a 2009 article, emphasise that structure, functions and ecosystem services are not identical or synonymous, and repropose the distinction into three types of services: the "intermediate" ones, understood as intermediaries between ecological functions and the spherulation (associated with structure and ecosystem processes), the "direct" or "final" ones, and the "benefits" or goods, the latter usually realised through human capital or technology (Fisher et al., 2009). As an example, considering the water resource, clean water is the result of an (intermediate) service generated by natural cycles of purification and storage (nutrient cycling) that occur independently of man and the nutrient cycle is therefore a service that man uses, but indirectly. Clean water, when consumed for drinking, is a benefit of ecosystem services. The supply of clean water is a (direct) service that man uses and requires capital and technological investments to obtain water for domestic use that is a benefit.

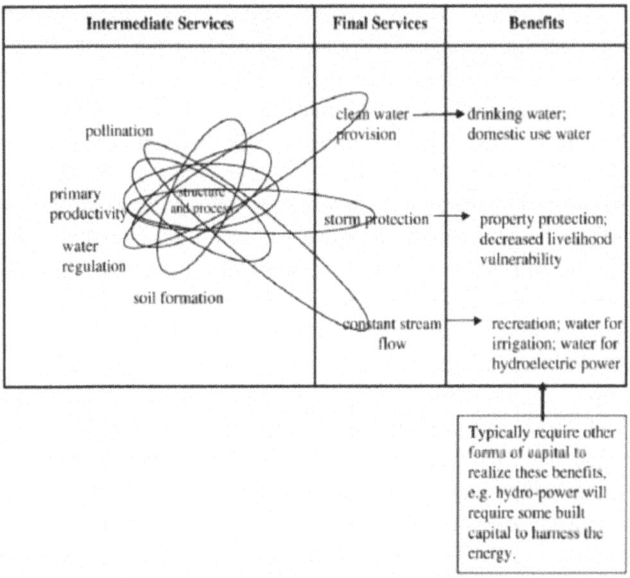

Fig. 3. Conceptual diagram of the relations between intermediate and final services, related to benefits (PTCP Province of Lecco) (Source: PTCP Province of Lecco - Monograph G - Ecosystem Services, https://www.provincia.lecco.it/pr-lecco-media/2022/08/03_Monograph_G_Services_Ecosystem_2022.pdf.).

The proposed classification scheme (Fig. 3), which divides ecosystem services into intermediate (e.g. pollination, soil formation, water regulation, primary productivity), final (e.g. water purification, storm protection, constant water flows) and benefits (e.g. drinking water, reduction of vulnerability levels, recreation, irrigation and hydropower production) is substantiated as being the most appropriate and allowing a differentiation between the former and the latter drinking water, reduction of vulnerability levels, recreation, irrigation water and for hydropower production, it is sublated to be the most appropriate and such that it is possible to differentiate whether it is the former (intermediate ES) or the latter (final ES), in relation to the connection with human benefit. With regard to the temporal and spatial dimension, however, ecosystems and ecosystem services are not equal in the terrestrial and marine landscape and are not static phenomena, but rather heterogeneous in space and evolving over time; spatio-temporal dynamics is indicated as a feature that can help understand and classify ecosystem services. For example, still with regard to water, its regulation guaranteed in the mountains by forest cover will become a benefit later on, in the form of regular and extensive water supply. Also with respect to landscape management, we note the relationship with the classification of ecosystem services on the basis of their spatial characteristics and the need to describe the relationships between the production of services and the places where the benefits accrue, traced to three situations: in situ, when services and benefits occur in the same place; omni-directional, when services are generated in one place and the benefits affect the surrounding territory; directional, when services give benefits in a specific place according to the direction of flow.

The distinction of services between supply and use gives the possibility, if the beneficiaries of the services are located in places other than where they are generated, to define compensations through the mechanism of payment for ecosystem services.

The different interpretations and systematisations of ecosystem services prompted the United Nations, under the UNEP programme (UN Environment Programme), and the EU, as the European Commission with the participation of several ministries from European countries, to support new research on the economics of ecosystems and biodiversity, which returns the results of the study on the economics of ecosystems and biodiversity, provides a new definition of ecosystem services and reformulates the classification of the same, implementing the method proposed by the MA.

The TEEB scheme identifies two blocks, that of 'Ecosystems and Biodiversity' and that of 'Human wellbeing (socio-cultural context)', which are linked in one flow direction by ecosystem services (Fig. 4). The first block includes 'biophysical processes and structures', which do not, such as, constitute a service for human beings, and the 'functions' or capacities that those processes possess. To the second block belong the 'benefits' for human beings, deriving from ecosystem functions, and the 'economic values' (monetary or non-monetary) that are attributed according to different methods of evaluation and quantification. The (ecosystem) services, placed between the two blocks, to be traced back to an anthropic dimension, are those that constitute the intermediary; these are declined according to the distinction into the four categories of supply, regulatory, habitat and cultural services. The definition of ecosystem services proposed by the TEEB group, "Direct or indirect contribution of ecosystems to human well-being", differs somewhat from that of the MEA in that the TEEB differentiates between services and benefits by specifying that the former can provide benefits in multiple and even indirect ways; ecosystems are, however, identified as a factor that makes a contribution to the well-being of the human population. The TEEB group proposes 22 types of eco-services, associated systemic services and grouped into one of the above-mentioned

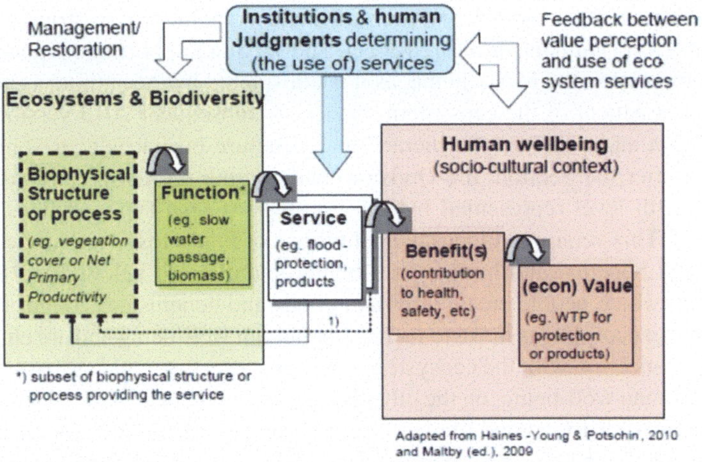

Fig. 4. First operational diagram, contained in the TEEB document, summarising the transition from ecosystem structure, ecosystem processes and human benefits (PTCP Province of Lecco) (Source: PTCP Province of Lecco - Monograph G - Ecosystem Services, https://www.provincia.lecco.it/pr-lecco-media/2022/08/03_Monograph_G_Services_Ecosystem_2022.pdf.).

four main categories; eliminating, compared to the MEA classification, the category of support services and adding, instead, the category of habitats, thus reinforcing the role of ecosystems in providing habitat for migratory species and as protectors of gene pools.

In 2011, a new classification was proposed by Roy H. Haines-Young and Marion B. Potschin, as an evolution of the MA categorisation, which was given the name CICES 'Common International Classification of Ecosystem Services'; the action that led to this new definition was promoted by the European Environmental Agency (EEA) with the aim of finding common grids to evaluate ecosystems in a uniform way at a European level (Haines, Potschin M.B. et alt, 2011). This classification has recently undergone revision, with the contribution of the System of Environmental-Economic Accounting (SEEA), under the United Nations Statistical Division (UNSD). The latest version of the CICES classification, V5.1 of 2018, provides the following definition of Ecosystem Services: "Ecosystem services are defined as the contributions that ecosystems make to human well-being. They are seen as arising from the interaction of biotic and abiotic processes, and refer specifically to the 'final' outputs or products from ecological systems" (Haines, Potschin, 2018). For the purposes of the ICESCR (International Covenant on Economic, Social, Cultural Rights) ecosystem services are thus defined as the contributions that ecosystems make to human well-being, with the clarification that these are seen as arising from the interactions of biotic and abiotic processes, and refer specifically to the 'final' outputs or products from ecological systems.

Going backwards to understand the evolution of the meaning of eco-systemic services according to the CICES classification, it can be seen that in the first CICES proposal, it identifies three levels, i.e.: that of the themes (later renamed sections), which includes supply, regulation, maintenance and cultural services; that of the classes (which, for example, with regard to supply includes food, materials, energy) and finally that of the groups (which, for example, with regard to food includes terrestrial, aquatic and marine plants, food animals, drinking water). The section Supporting (support services) is elided, compared to the MA scheme, as it is considered by the authors as not relevant to the final assessment of the ecosystem service. A subsequent 2013 version takes the original version and confirms the hierarchical structure but introduces some changes by inserting, after the Section, the Division, anticipating the Group and adding, after the Class, a fifth level represented by the Typology (Class Type) (Haines, Potschin, 2013) (Fig. 6). This version provides the definitions of final ecosystem services, ecosystem goods and benefits, and human well-being, summarised below. Final ecosystem services, from which people create or obtain goods and benefits, are identified with the contribution that ecosystems make to increasing human well-being and are characterised by having a connection with the ecosystem functions, processes and structures that generate them. Human well-being, on the other hand, is that which provides adequate access to the basic materials for a good life, necessary to sustain freedom of choice and action, health, good social relations, and security; the conditions for such well-being depend on the overall supply of ecosystem goods and benefits, the provision of which can change the state of well-being. The three strands of ecosystem services are: provisioning services, which include all food, materials and energy from living systems; regulating and maintenance, which include all the mechanisms of living organisms that can regulate the environment and affect the human condition; and cultural services, which include all the non-material and normally non-consumable ecosystem services that affect people's physical and mental conditions. The cascade model (Fig. 5), in the latest formulation described by Potschin and Haines-Young, is the one depicted in the figure below, in which the (ecosystem) services are placed in the 'final services' box (Haines, Potschin et al., 2016).

In 2012 The MAES, a working group established by the European Commission, with the expanded name of "Mapping and Assessment of the Ecosystem and their Services", set out to support and increase the mapping of ecosystems and their services for the European territory(Fig. 7).

Four interim reports on specific aspects are published under this project (in 2013, 2014, 2016a, 2016b) and the final report, 'Mapping and Assessment of Ecosystems and their Services: An analytical framework for ecosystem condition,' is published in 2018 by the European Union (Maes J et, 2018). In addition, a specific report on soil ecosystems is published in 2018 and the sixth report entitled 'Natural Capital Accounting: Overview and Progress in the European Union' is published in 2019 (EU, 2019).

In 2018, MAES, on behalf of the European Commission, DG Environment, published the report 'Mapping and Assessment of Ecosystems and their Services Soil ecosystems' in which ecosystem services specifically related to soil are considered, again for the purpose of mapping and assessing the condition of soil ecosystems and their services, identifying reference indicators. In the foreword to the document, it is pointed out that soil ecosystem services can contribute to the achievement of the United Nations Sustainable Development Goals (SDGs) and in particular Target 15.3 concerning combating desertification and restoring degraded soils and lands. The report identifies, in illustrative tables, the soil ecosystem services pertaining to the three main categories, identifying the functional processes required to generate these services and provides, again in tables,

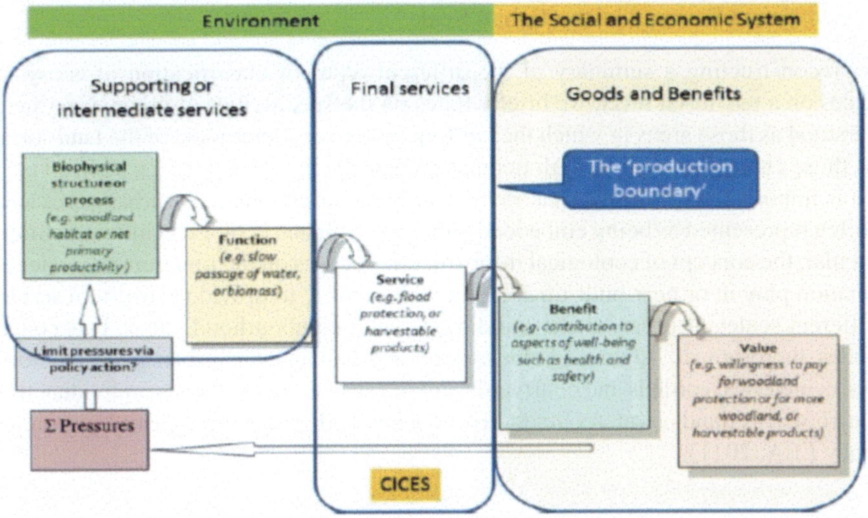

Fig. 5. Ecosystem Services cascade model, Haines, Potschin et al., 2016 (PTCP Province of Lecco) (Source: PTCP Province of Lecco - Monograph G - Ecosystem Services, https://www.provincia.lecco.it/pr-lecco-media/2022/08/03_Monograph_G_Services_Ecosystem_2022.pdf.).

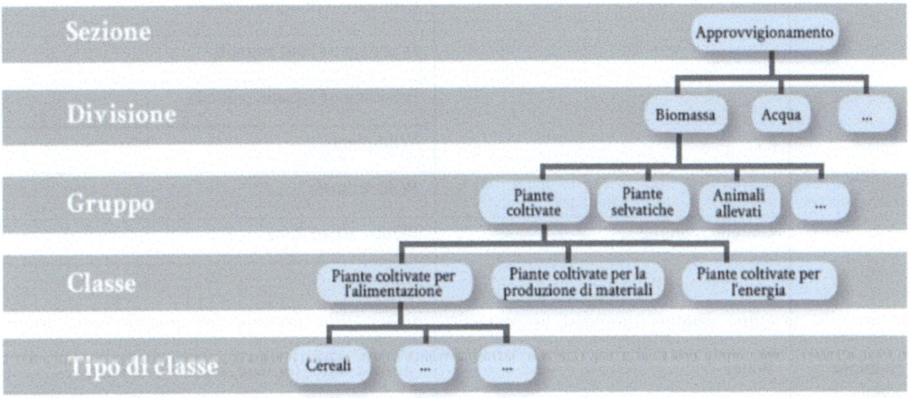

Fig. 6. Latest version of the CICES confirming the above-mentioned five-level articulation, represented in the following diagram (PTCP Province of Lecco) (Source: PTCP Province of Lecco - Monograph G - Ecosystem Services, https://www.provincia.lecco.it/pr-lecco-media/2022/08/03_Monograph_G_Services_Ecosystem_2022.pdf.).

for each ecosystem service, the list of related indicators for which the spatial dimension of reference (regional or local) for their application is specified, whether they are supply or use services, and the availability of data. Indicators for assessing soil pressures and ecosystem conditions are also listed in the document (Van der Meulen S., Maring L., 2018).

2.2 Ecosystem Services at the Urban Scale

After reconstructing a summary of the different types of classification of ecosystem services on a territorial level, we briefly focus on the specificities of urban ecosystems, understood as those areas in which the built-up system occupies most of the land surface or as those characterised by a high population density.

It is important to point out that, in the context of urban planning, urban ecosystems are often represented as being embedded with both built and ecological infrastructure. In particular, the concept of ecological infrastructure focuses on the function that water and vegetation play in or near built-up areas and their ability to provide ecosystem services at different scales of urban space (building, street, neighbourhood, city). This concept includes, therefore, all 'green and blue spaces' in urban and peri-urban areas, including: parks, cemeteries, gardens and courtyards, urban gardens, urban forests, individual trees, green roofs, wetlands, watercourses, rivers, lakes and ponds (European Environment Agency EEA, 2011).

CICES - Version 5.1 - Structure		
Sector	***Division***	***Group***
Biotic supply	Biomass	Cultivation of terrestrial plants
		Cultivation of aquatic plants
		Breeding of land animals
		Breeding of aquatic animals
		Wild plants
		Wild animals
	Genetic material	Plants, algae, fungi
		Animals
	Other supplies from biotic resources	More
Biobiotic supply	Water	Surface water for feed, materials or energy
		Groundwater for food, materials or energy
		Other supplies from aquatic ecosystems
	Supplies from non-aquatic natural abiotic ecosystems	Mineral substances for food, materials and energy
		Non-mineral substances or properties of ecosystems used for food, materials or energy
		Other minerals or non-mineral substances or ecosystem properties used for food, materials or energy

Fig. 7. CICES Ecosystem Services Classification Scheme (PTCP Province of Lecco) (Source: PTCP Province of Lecco - Monograph G - Ecosystem Services, https://www.provincia.lecco.it/pr-lecco-media/2022/08/03_Monograph_G_Services_Ecosystem_2022.pdf.).

CICES - Version 5.1 - Structure		
Sector	***Division***	***Group***
Adjustment and Mainte-nance biotics	Transformations of biochemical or physical inputs of ecosisthemes	Assimilation of waste or t o x i c substances by living processes
		Assimilation of disorders of antr
	Regulation of physical, chemical and bi-ological conditions	Flow regulation and east events
		Maintaining life cycles and pro gene banks
		Pest and disaster control
		Soil quality regulation
		Water conditions
		Composition and conditions of the atmosf
	Other regulation and maintenance by processinations	More
Adjustment and Mainte-nance abio-tics	Transformations of biochemical or physical inputs of ecosisthemes	Assimilation of waste or t o x i c substances by non-living processes
		Assimilation of disorders of antr
	Regulation of physical, chemical and bi-ological conditions	Flow regulation and east events
		Maintaining physical conditions, ch
	Other regulation and maintenance by abiotic processes	More

CICES - Version 5.1 - Structure		
Sector	***Division***	***Group***
Cultural bio-tics	Direct, on-site or external interactions with living systems depend on the pres-ence of environmental contexts	Physical and experiential interactions co
		Intellectual interactions and natural representa-tion
	Indirect, remote, indoor interactions with sistemivents that do not require presence in environmental counts	Spiritual, symbolic or natural environment inter-actions
		Other biotic characteristics
	Other characteristics of living systems that have cultural unsignification	More
Cultural abio-tics	Direct on-site or external interactions with natural physical systems that de-pend on the presence of environmental contexts	Physical and c abiotic experiential interactions of the natural environment
		Intellectual interactions and abiotic representa-tions of the natural environment
	Indirect, remote, indoor interactions with system physics that do not require presence in environmental contexts	Spiritual, symbolic or abiotic interactions of the environment n
		Other abiotic characteristics that h do not use
	Other abiotic features of nature that have cultural significance	More

Fig. 7. *(continued)*

Defining clear-cut boundaries for urban ecosystems is, very often, not an easy task, as many of the important flows and interactions essential for understanding the functioning of urban ecosystems expand far beyond urban boundaries that can be traced back to administrative or biophysical reasons. For this reason, the scope of cataloguing and investigating urban ecosystems must be considered beyond the canonical urban perimeters, as it encompasses not only the ecological infrastructure within the city, but also the hinterland - including catchment areas, agricultural areas and wooded areas - that are explicitly affected by flows from the urban core.

In recent years, a large body of literature has improved our knowledge and understanding of urban ecosystem services, their role and their biophysical, economic and socio-cultural components. Furthermore, urban ecosystem services have been the subject of consideration in the aforementioned major projects such as the *Millennium Ecosystem Assessment* (MA, 2005) and *The Economics of Ecosystems and Biodiversity* (TEEB, 2011) and, in recent years, have received increasing attention within the more general debate on ecological infrastructure. However, despite the fact that more than half of the world's population now lives in cities, the interest assigned to urban ecosystems in the literature on ecosystem services is still quite limited compared to that assigned to other spatial-level ecosystems.

The following is a classification and description of ecosystem services in urban areas, focusing on how they can contribute to increasing the quality of life in cities. Building on previous categorisations of ecosystem services (Daily 1997; et al.), the aforementioned *Millennium Ecosystem Assessment* (MA 2005) and *The Economics of Ecosystems and Biodiversity* (TEEB 2011) have grouped ecosystem services into four main categories: supply or provisioning, regulating, cultural and recreational, supporting or habitat.

Supply services include all material products obtained from ecosystems, including genetic resources, food and fresh water. *Regulation services* include all benefits obtained from the regulation of eco-systemic processes, including the regulation of climate, water and some human diseases. *Cultural services* are the non-material benefits that people obtain from ecosystems through spiritual enrichment, cognitive development, recreation and aesthetic experience, as well as their role in sustaining knowledge systems, social relations and aesthetics. *Supporting or habitat services* are those necessary for the production of all other ecosystem services. Examples include biomass production, nutrient cycling, water cycling, provision of habitat for species and maintenance of gene pools and evolutionary processes.

Since different habitats provide different types of ecosystem services, general classifications must be adapted to specific types of ecosystems. Urban ecosystems are particularly important in providing services that have a direct impact on human health and safety, such as air purification, noise reduction, urban cooling and water runoff mitigation.

3 The European Blue Green City Project

In the course of the research underlying this contribution, a number of urban planning projects were analysed that aim to integrate ecological services into the plan in order to strengthen the resilience of urban areas and increase their bio-diversity. Among the many projects examined - the urban regeneration of disused industrial areas of the Municipality of Trento, the Territorial Plan of Milan, the Municipal Plan for the City of Mangalia, the Territorial Plan of the City of Lecco, etc. - the Blue Green project is examined.

The Blue Green Cities project, which concerns a group of municipalities belonging to the Piemonte region, is chosen for the rigorous methodological approach developed to outline the planning actions starting from the mapping of the ecosystem SEs in the various UPAs (Landscape and Environmental Units) into which the entire area of intervention is subdivided and to constitute; with the drafting of the Guidelines for Green and Blue Infrastructures that accompany the plan, this project represents a capitalised experience that, if well understood, can be translated and adapted to other territorial contexts.

The BLUE GREEN CITY project ("*Green and Blue Infrastructure for sustainable cities*") was activated under the Interreg Europe Programme 2014–2020 building on the results of an earlier applied project LOS_DAMA! (Green Infrastructure for Better Living) financed by the Alpine Space Programme 2014–2020. With the pilot activity of this project, a methodology for the sustainable planning of Green and Blue Infrastructures and a wide area planning scheme were defined, with an approach that integrates landscape matrices and the mapping of potential ecosystem services with the aim of providing strategic responses to needs and vulnerabilities and enhancing the resilience of the study area through the planning of green and blue infrastructures.

Subsequently, with the Blue Green City project, the methodology and the wide area planning scheme (Figs. 8 and Fig. 9) were tested and validated in a more restricted area involving the Union of Municipalities North East of Turin and the Municipality of Mappano in a participatory planning process that led to the definition of a more detailed inter-municipal planning scheme, with the identification of actions and strategies that can be implemented on a local scale and the definition of specific guidelines. The result is a comprehensive guide for the strategic planning, design and sustainable maintenance of green and blue infrastructure with an integrated, participatory and place-based approach, addressed to local public administrations, technicians and stakeholders.

In the *Blue Green City* project, the Intermunicipal Planning Scheme, comprising the Local Action Programme (LAP), integrates the following objectives into the spatial planning and governance tools to promote collaboration and multilevel exchanges to define innovative planning models oriented towards sustainability and increasing the resilience of territories; to improve policies that promote Green and Blue Infrastructures as an integral part of a local or regional strategy for the conservation of Natural Capital; to increase the knowledge of stakeholders on the concepts of ecosystem services and the value of Green and Blue Infrastructures; to build a programme of actions to increase the capacity of territories to adapt and respond spontaneously to environmental, social and economic changes.

The starting assumption of the project is to consider it necessary to construct an interpretative diagnostic system to effectively describe the structuring characteristics of the various Spatial Units that make up the Pilot Area and the problems and values from which to extract the priority Ecosystem Services at different scales. The planning process developed for the project is develops in five phases. The first phase consists of building a framework of basic knowledge in order to understand the structure and organisation of the landscape at different scales. It consists of three stages: general knowledge

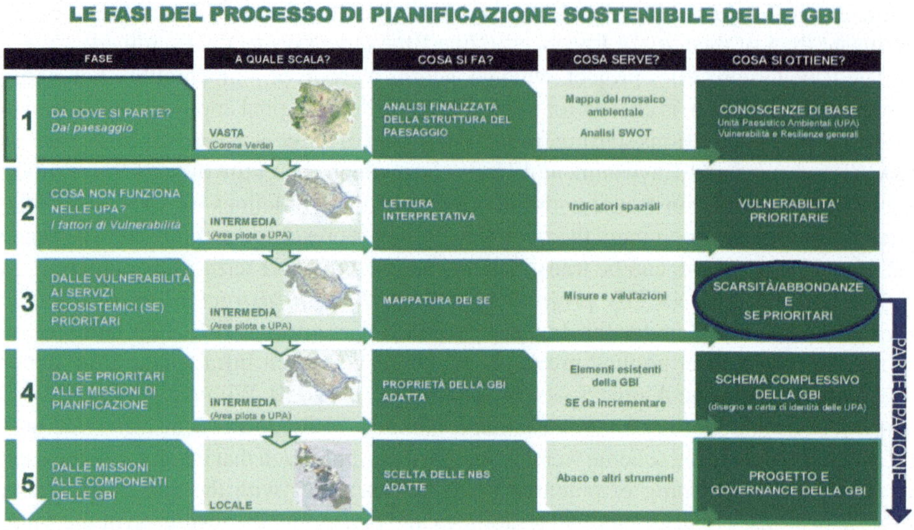

Fig. 8. Logical diagram of the sustainable planning process (Gibelli et al., 2022) (Source: Project Blue Green City, Green and Blue Infrastructure strategically planned Guidelines in https://www.regione.piemonte.it/web/sites/default/files/media/docu-menti/2023-01/Line%20Guide_complete.pdf.)

Fig. 9. The logical scheme of the Sustainable Planning Process (Gibelli et al., 2022) (Source: Project Blue Green City, Green and Blue Infrastructure strategically planned Guidelines in https://www.regione.piemonte.it/web/sites/default/files/media/docu-menti/2023-01/Line%20Guide_complete.pdf.)

to understand the structure of the landscape, landscape-environmental mapping to represent the diversity of the territorial area under examination, (the "Green Crown"), to identify structural patterns and subdivision into sub-systems, (UPA Territorial Units of Reference), and finally the selection of reading keys suitable for the definition of needs and their localisation in order to be able to transform them into policies and actions, corresponding with the concepts of Vulnerability and Resilience. The second phase consists of identifying priority 'needs' to work on through spatial indicators that estimate the vulnerability of socio-ecological systems; this is indispensable since not all vulnerabilities are primary. The latter are those that most affect the equilibrium of spatial units and thus the possibilities for spontaneous adaptation. They generally belong to both the large and intermediate scales and can be defined through spatial indicators and the classification of results in the different spatial units. After defining Vulnerability and Resilience on the large scale, one then proceeds to the estimation of vulnerability on the intermediate scale, i.e. the pilot area and the 14 UPAs. Finally, by comparing the vulnerability factors of the UPAs with those of the broad scale (Green Crown), it is possible to understand which broad scale processes affect the Vulnerability of the pilot area and vice versa, and those vulnerabilities that correspond between the intermediate and the broad scale are considered priorities. Once the priorities have been identified, priority ecosystem services are defined, i.e. those effective in resolving or mitigating the various PAUs whose needs they signal. The third phase concludes with the definition of scarcity and abundance levels of priority ecosystem services within the various PAUs. In the fourth phase, functional "Planning Missions" are developed to enrich the scarce ecosystem services through the definition of the multifunctional Green and Blue Infrastructure (IVB), in which the necessary functions are clearly dictated by the priority ecosystem services (Gibelli, D'Ambrogi, 2022); these "Planning Missions", organised into Objectives-Strategies and Actions, are reported in the Summary Sheets (which constitute a sort of identity card for the UPAs). Below is a summary sheet referring to a specific UPA (Fig. 10).

UPA	PERCHE'	COSA	DOVE	
	Obiettivi	Missioni di Pianificazione	Paesaggi	Sistemi funzionali
Si riporta l'UPA in esame	Si riportano uno o più obiettivi specifici della UPA	Sono descritte le azioni che attuano gli obiettivi. Ogni obiettivo può prevedere una o più azioni	Si individuano le categorie di paesaggi per i quali quelle determinate azioni sono efficaci	Le categorie di paesaggi sono eventualmente specificate ulteriormente in sistemi funzionali
La riga seguente riporta un esempio di scheda compilata tratta dal progetto LOS_DAMAI				
6 Alta pianura	Ridurre la conflittualità tra elementi incompatibili	Mitigare le infrastrutture, con interventi attenti anche alla ricucitura del paesaggio originario e non solo alla mitigazione visiva	Paesaggio delle infrastrutture	Agroecosistema Ecosistema Urbano
		Lavorare sui margini degli insediamenti sparsi per ridurre le interferenze con le aree agricole; impedire/ ridurre la frammentazione delle aree agricole	Paesaggi periurbani rurali	Ecosistema Urbano/Agroecosistema

Fig. 10. Overview diagram of a UPA specification (Gibelli et al., 2022) (Source: Project Blue Green City, Green and Blue Infrastructure strategically planned Guidelines in https://www.regione.piemonte.it/web/sites/default/files/media/docu-menti/2023-01/Line%20Guide_complete.pdf.)

The "Actions" or "Planning Missions", to be implemented by means of the most suitable Nature Based Solutions (NBS) and policies, which outline the most effective infrastructures to erect Ecosystem Services are translated into Planning Schemes. The scheme elaborated with the European projects Los_Dama! and Blue Green City is the response to the needs that emerged from the analyses and assessments carried out at the different scales considered and specifically: at the intermediate scale it was developed within the Los_Dama! project and represents an overall IVB development scenario aimed at increasing the provision of priority Ecosystem Services, proposing a stock of appropriate, feasible and sustainable actions and solutions capable of attracting the necessary funding (the Planning Missions); at the local scale it was developed within the Blue Green City project for the Unione dei Comuni Nord Est Torino and the municipality of Mappano and represents a programme of actions aimed at implementing the Planning Missions through the implementation of effective actions starting from the priority ones selected to respond to the priority vulnerabilities and selected during the participatory process that accompanied the whole planning process.

The inter-municipal planning scheme (envisaged by the Blu Green City project), of which the Programme of Actions constitutes the heart, is composed of the mapping of the Green and Blue Infrastructure at present on which to graft the overall planning scenario of the same Infrastructure, the map of the IVB project, which represents the strategic areas (the areas of intervention) where to build the new IVBs and the Nature Based Solutions for the completion of the network, the actions to start implementing the scenario (the Programme of Actions) and the priority action sheets that contain the information and data needed to develop them. In particular, the Programme of Actions (PAL) contains those actions that prioritise the territorial problems and are feasible because they are agreed and shared with the stakeholders, the actors involved are defined, they are accompanied by a schedule, the timeframe and the approximate costs of realisation. The fifth phase of the project's methodological pathway concludes the planning process at the local level with the selection of the most suitable NBSs to increase priority ecosystem services; the NBSs are therefore the ultimate result of the planning process and the local response to landscape needs. The selection of the most effective NBSs represents the final building block for the construction of the IVBs; in this sense, they come downstream of all the previous steps and incorporate them in terms of a concrete response to landscape needs.

The planning process just illustrated allows for the setting up of planning tools suitable for responding effectively to real adaptation needs of territorial systems at different spatial scales. The plan is, therefore, supported by a "dynamic frame of reference" within which ecosystem services become tools to reduce vulnerabilities and improve the resilience of PAUs and the whole territory in a multi-scalar approach (Gibelli et al., 2022).

4 Conclusions

In recent decades, scientific evidence of the benefits of natural capital and biodiversity has been consolidated; these environmental and social benefits that constitute Ecosystem Services make cities and societies more sustainable. This diffuse naturalness is mostly constituted by green and blue infrastructures and by Nature Based Solutions (NBS), which have recently taken on new and central roles in urban and territorial planning, no longer contributing exclusively in quantitative terms (town planning standards Ministerial Decree No. 1444/1968) as a compendium accompanying urban development, but above all as strategic 'assets' for the sustainability of urban settlements and the quality of human life in the city (ASVIS, 2023). In this context, the European project Blu Green City is certainly a good urban planning practice for the methodological approach used in the valorisation process of the Green and Blue Infrastructure aimed at the production of ecosystem services in the awareness of having to consider them and the permeable soils that support them as the pivot of a new season of urban and territorial planning for the ecological transition of the city (ASVIS, 2023).

The knowledge of the contents and the method characterising the Blue Green City project gave rise to some important general reflections, reported below, on the IVBs, SEs, NBSs and the close concatenation between these and planning actions, which certainly go beyond the specific case dealt with, but which can instead contribute to building a wealth of knowledge useful in similar conditions, even if in different territorial contexts, in which the objective of increasing the sustainability of cities and territories is achieved through effective planning tools.

IVBs are created to produce environmental, social and economic benefits for the territory and its communities, so before planning, it is necessary to be clear on which benefits to aim for.

IVBs are governance tools for adaptation and for improving the sustainability of policies and plans, so design is only one part of the project. Participation, integrated policies and monitoring of benefits are the indispensable tools for action.

The design of IVBs must express the community's "visionary response" to the demands of environmental, social and economic adaptation and sustainability; it cannot therefore be a specific design, but rather a mosaic of localised demands that can be resolved through suitable NBSs that constitute a system capable of enhancing ecological functions and thus Ecosystem Services.

The Planning Scheme, which concludes the methodological path proposed by the Los Dama! and Blu Green City projects, represents an effective tool for increasing the sustainability of cities and territories by offering a response to the challenges that climate, social, environmental and economic changes are also throwing at planning. It is an active tool for an overall redevelopment of the environmental landscape system capable of responding to the contemporary needs for resilience and increased sustainability of both natural and cultural landscapes. It constitutes a strategic framework for local planning and territorial transformations; it is a catalyst for economic resources that can be allocated to the landscape-environmental redevelopment of the territory through the construction of the Green and Blue infrastructure; it stands as a useful tool for the knowledge of the landscape and its elements, contributing to the awareness of the population towards the great themes of global changes.

The proposed planning process is characterised by providing a comprehensive vision of the landscape/environmental system that goes beyond, at least in part, the more traditional approaches.

It is not a closed plan to be realised with public economic resources, but a process in which several actors can participate, without prejudice to the objectives, expected results and certain fundamental principles and criteria. The planning process does not end with the realisation of the project; the IVBs and the spaces they are made up of are living, dynamic organisms, which must modify themselves over time, adapting, mutually with the cities, to changes.

Finally, it is necessary to monitor and measure, using spatial indicators used for Vulnerability and Resilience, what is being achieved in order to understand whether the objectives are being met, in particular that of improving the health of landscapes, responding to vulnerabilities and increasing Ecosystem Services, and also which actions have been most effective.

In light of what emerged in the analysis of the Blue Green Cities project, several useful elements were transferred to the implementation of the objectives of Action 1, Pilot Project 4.6.1 of the Spoke 4 of the Tech4You Innovation Ecosystem and, in particular, the recognition of the centrality of the Green and Blue Infrastructures, of which all the constituent and applicative aspects have been examined, in the process of valorising ecosystem services and safeguarding biodiversity and, again, the identification of a possible way forward, also in the case of Pilot Project 4.6.1, to assess the vulnerability and resilience of a territory through its capacity to provide ecosystem services and to be able to direct planning actions accordingly.

Acknowledgments. This research work is the result of the research activities conducted within the following PNRR research project funded by the European Union – NextGenerationEU: "Pilot Project 4.6.1 "Climate adaptation plans for the reduction of the ecological footprint and ecological debt, aimed at improving the conservation and transformation response in terms of resilience and quality of life in the Calabrian and Lucanian urban and territorial systems" (Goal 4.6 - Planning for Climate Change to boost cultural and natural heritage: demand-oriented ecosystem services based on enabling ICT and AI technologies - Tech4You Innovation Ecosystem), Finanziato dall'Unione europea- Next Generation EU, Missione 4 - Componente 2 - CUP C33C22000290006 (ECS_00000009).

References

Rawtorth, K.: A safe and just space for humanity, Oxfam Discussion Paper (2012)

Gibelli, G., et al.: Green & blue strategically planned infrastructure - guidelines. Piedmont Region (2022)

Caldarice, O., Salata, S.: Assessing ecosystem services in the plan as a response to territorial vulnerability: a methodological reflection starting from the proposed land consumption law. In: Piedmont in Values and Assessments, vol. 22 (2019)

Santolini, R., Morri, E.: Ecological criteria for the introduction of evaluation and remuneration systems for ecosystem services (ES) in design and planning. In: Arcidiacono, A., Di Simine, D., Oliva, F., Ronchi, S., Salata, S. (eds.) La dimensione europea del consumo di suolo e le politiche nazionali, pp.149–154. INU edizioni srl, Roma (2017)

PTCP Province of Lecco - Monograph G - Ecosystem Services. https://www.provincia.lecco.it/pr-lecco-media/2022/08/03_Monograph_G_Services_Eco-systemic_2022.pdf

Ehrlich, P.R., Ehrlich, A.: Extinction: The Causes and Consequences of the Disappearance of Species'. Random House, New York (1981)

Costanza, R., et al.: The value of the world's ecosystem services and natural capital. Nature **387**, 254–260 (1997)

Costanza, R., et al.: Changes in the global value of ecosystem services. Glob. Environ. Chang. **26**, 152–158 (2014)

Daily, G.C.: Nature's Services: Societal dependence on Natural Ecosystem. Island Press (1997)
Millennium Ecosystem Assessment: Ecosystem and Human Wellbeing: Synthesis. Island press, Washington, DC, World Resources Institute (2005)
Fisher, B., et al.: Defining and classifying ecosystem services for decision making. Ecol. Econ. **68** (2009)
Pavan, S., et al.: The Economics of Ecosystem and Biodiversity: The Ecological and Economic Foundation, edited by Pushpam Kumar. Earthscan, London and Washington (2010)
Haines-Young, R.H., Potschin, M.B.: Ecosystem service: exploring a geographical perspective. In: Progress in Physical Geography (2011)
Haines Youg, R.H., Potschin, M.B.: Common international classification of ecosystem services (CICES) V5. 1 and guidance on the application of the revised structure. Nottingham, Fabis Consulting, Ltd. (2018)
Haines-Yong, R.H., Potschin, M.B.: CICES V4.3 - revised report prepared following consultation on 465 CICES Version 4, August-December 2012. EAA (2013)
Haines-Young R.H., Potschin M.B.: Defining and Measuring Ecosystem Services. Routledge Handbook of Ecosystem Services, pp. 25–44 (2016)
Maes, J., et al.: Mapping and Assessment of Ecosystems and Their Services: An Analytical Framework for Ecosystem Condition. Publication office of the European Union, Luxembourg (2018)
EU: Natural Capital Accounting: Overview and Progress in the European Union, 6th Report, Publication office of the European Union (2019)
Van der Meule, S., Maring, L.: Mapping and assessment of ecosystems and their services soil ecosystems, report no.1.2, EU DG Environment (2018)
European Environmental Agency: Annual Report 2011 (2011)
TEEB: TEEB Manual for Cities: Ecosystem Services in Urban Management. The Economics of Ecosystems and Biodiversity (TEEB). Geneva (2011)
Gibelli, G., D'Ambrogi, S.: Ecosystem services in planning processes. Ecoscienza **2**, 26–28 (2022)
LOS_ DAMA! Landscape and Open Spaces Development. https://www.Alpine. Metropolitan Areas. https://www.alpine-space.eu/project/los_dama/
Blue Green City. https://www.interregeurope.eu/bluegreencity/
Asvis: Report 2023. https://Asvis.it/Rapporto-2023
Project Blue Green City: Green and Blue Infrastructure Strategically Planned Guidelines. https://www.regione.piemonte.it/web/sites/default/files/media/documents/2023-01/Linee%20Guida_completa.pdf

Ecosystem Services Innovation and Integration of Knowledge for the Contemporary City

Natalina Carrà(✉)

DArTe Department - Mediterranea University, Reggio Calabria, Italy
ncarra@unirc.it

Abstract. The territory, today, is very vulnerable and unprepared for possible changes linked to the uncontrolled exploitation of resources which causes land consumption. Cities increasingly suffer impacts, often catastrophic, because they have to deal with climate change. Contemporary urban planning is therefore urged to search for new paradigms more suited to the scenario of malaise in which we live. Specifically, the role of the evaluation of Ecosystem Services (ES) is of particular importance as a cognitive and interpretative support for the different functions of the soil, directly connected to human well-being and their value for the quality of life. Furthermore, the potential of environmental infrastructures as a tool for structuring a widespread and sustainable project for cities is fundamental, capable of enhancing the different vocations of the urban environment, starting from the conservation and enhancement of the ecosystem functions that the soil provides. The paper highlights the importance of the function of valorising ES as a possible response to the needs of the contemporary city, through new scientific methodologies, which offer a possible approach to the prevention of ecological problems caused by human action, and to the resolution of conflicts arising from changes in land use.

Keywords: Ecosystem Services · Innovation · Contemporary City

1 Ecosystem Services and the Contemporary City

Introduction

The cities of the third millennium must deal with epochal changes: those linked to climate, pollution, land consumption and energy consumption, attempting to increase the ability of the urban environment to deal with the effects of these changes.

Land consumption has made cities more fragile in the face of the devastating effects of climate change. A paradigm shift is therefore necessary. Territorial planning must protect the soil and related ecosystem services (ES); it must also promote the recovery and restoration of those soil functions which in the urban context can help implement innovative ES integration strategies in favor of and in response to climate adaptation.

All soil ES are important, and in the urban environment some are particularly important, such as those that can increase resilience to climate change. Awareness has grown on issues such as the fight against land consumption and urban regeneration. The effects

F. Moraci et al. (Eds.): NMP 2024, LNNS 1272, pp. 128–137, 2025.
https://doi.org/10.1007/978-3-031-82927-7_8

of climate change make it increasingly clear that the urban development model implemented to date is no longer sustainable. On the one hand, we need to plan to prevent a further worsening of the impacts of these effects on urban centers by limiting land consumption. On the other hand, at the same time, we must act to implement all the corrective measures necessary to mitigate these effects.

The functionality of ecosystems and the interactions that man develops with them can determine the overcoming of these difficulties and problems. This innovative approach to urban planning represents a possible and valid response to the increasingly widespread and worrying soil sealing, which is the main cause of the loss of biodiversity and the destruction of both rural and natural landscapes.

In this sense, if the evaluation and mapping of ES, providing qualitative knowledge of the soil's functionality, can direct the transformations of the territory from a perspective of ecological sustainability, on the other hand environmental infrastructures constitute a methodological approach to urban planning, currently at the center of the main Italian and international landscape and territorial planning experiences, which addresses the project of urban and peri-urban open spaces, in a trans-scalar perspective, with the aim on the one hand of enhancing existing ES and on the other of create, through NBS (Nature-Based Solutions) interventions, an urban structure that is resilient and adaptive to environmental pressures and more generally to climate change, in which the provision of natural spaces also provides a possible response to conditions of safe use of open spaces urban even in situations of new emergencies.

1.1 The Urban Ecosystem

In the seventies of the last century, following the studies on natural ecosystems initiated by the IBP (International Biological Programme)[1], later replaced by the Man and Biosphere program[2], scholars realized that they had to also consider man as part of the ecosystem calculation. of nature, and its main user and modifier. The main innovation was precisely that of considering the city as an *ecosystem*, that is, characterized by a constant flow of incoming and outgoing matter and energy. The city, therefore, considered as an ecosystem, but with the difference that a natural ecosystem is capable of self-feeding with a final budget in balance while the city lives *«crossed by a flow of materials, of atmospheric gases, of water, means of transport, people, and expels, like any living organism, the waste of its metabolism towards the air, rivers, sea, soil, the urban environment itself and its inhabitants, towards the surrounding environments»*[3].

The urban ecosystem[4] can therefore be assimilated to a highly complex system, which depends on the levels of anthropization and social and technological development to satisfy the needs of the population. The city, therefore, is not only the place where human beings coexist, but it is a complex ecosystem, which arises from man's need to fit in, as a social animal, in an environment (the urban one) made up of equally complex, between people, between people and nature and between people and cities. Paul Crutzen,

---

[1] https://www.nasonline.org/about-nas/history/archives/collections/ibp-1964-1974-1.html.
[2] https://www.mase.gov.it/pagina/il-programma-uomo-E-biosfera-mab.
[3] G. Nebbia, 2006.
[4] M. Nicoletti,1978.

Dutch Nobel Prize winner, defines this era as the "Anthropocene"[5] to underline that it is Homo Sapiens who play a main role in every ecosystem on this planet.

Therefore a new era, characterized mainly by the action of man, which has effects not only on the atmosphere and climate, but on all the cyclical processes of transformation of the planet.

Here, the cities of the third millennium must, therefore, deal with epochal changes caused mainly by human action: those linked to climate, pollution, land consumption and energy consumption, and therefore find themselves having to attempt to increase the capacity of the urban environment to deal with the effects caused by these changes. Starting from land consumption, which has made cities more fragile in the face of the disastrous effects of climate change. Soil is a precious and non-renewable resource: consumption and waterproofing lead to the loss of the numerous ES that the soil offers.

Planning for climate adaptation and the protection and restoration of soil and related ES are closely related. Together with urban planning standards and urban greenery, ES are considered structuring elements for the contemporary city due to their benefits which do not exclusively concern the large or global scale, but also the local one, as they act positively on the well-being of the community and the public health. ES are provided by natural soils and are often compared, in the contemporary Italian debate, with urban planning standards, *but "they depend on the complexity relationships that develop in the organization of ecological variables and cannot be standardized in an undifferentiated parameter"*[6].

1.2 The Functions of Ecosystem Services

The concept of ES[7] is the core of the process of valorisation of natural capital and a "strong" reason for the conservation of nature and biodiversity. In systemic terms, these services can be considered as flows provided by stocks of natural capital, and a large part of them are indispensable for the life of man and nature itself.

The National Biodiversity Strategy for 2030 is based on the vision that *"Biodiversity and ecosystem services, our natural capital, are conserved, evaluated and, as far as possible, restored, for their intrinsic value and so that they can continue to sustain sustainable economic prosperity and human well-being despite the profound changes taking place at global and local levels"*[8].

In fact, the large quantity and variety of functions and SEs that the soil has are specifically ecological functions that have a direct and indirect benefit for humans, in close relation with the conditions of the animal and plant communities that support them.

The concept of ES has been subject to different and often contradictory definitions and interpretations: sometimes it is used to describe the internal functioning of the

[5] Crutzen, P.J. (2005)

[6] C. Giaimo, (2019).

[7] The expression SE was introduced for the first time by Robert Costanza in 1997: Ecosystem services consist of flow of materials, energy, and information from natural capital stocks which combine with manufactured and human capital services to produce human welfare.

[8] https://www.mase.gov.it/sites/default/files/archivio/allegati/biodiversita/strategia_nazionale_biodiversita_2030.pdf.

ecosystem (energy flows, interactions of the food network) and sometimes it refers to the benefits deriving from ecosystem processes at man (food production). According to the definition given by the Millenium Ecosystem Assessment, ES are those *"multiple benefits provided by ecosystems to mankind"*. These services, therefore, have a close relationship with the well-being conditions of the community.

The current conditions of global warming, the increase in heat islands, the risk of flooding and more have sent a serious alarm to all those skills that deal with the territory, its maintenance, and its transformations. The current territory is vulnerable and unprepared for possible changes, also due to uncontrolled land consumption. It is therefore the cities that suffer impacts, often catastrophic. The urbanization processes that have occurred in recent decades have generated a series of negative impacts on the environmental balance of the territories, resulting in the irreversible transformation of permeable natural surfaces, compromising the ecological functions of the soils and the loss of their important ES with important environmental and socio-economic impacts for the community. On an urban and peri-urban scale, the progressive infrastructure of the territory has produced the fragmentation of natural capital and the reduction of large areas of naturalness. The report *"Land consumption in Italy 2023"*, published by ISPRA[9], confirms that land consumption, which has reached a speed of 2.4 square meters per second, has advanced, in just twelve months, by another 77 km^2: over 10% more than in 2021. This consumption had not slowed down even in 2020, despite the months of blocking activities during the lockdown.

Land consumption, therefore, generates the loss of a plurality of precious ES. To understand the importance of soil and its role in maintaining a balanced and healthy ecosystem, it is necessary to know what its ES are (i.e. the processes through which natural ecosystems satisfy the needs of human beings and contribute to their well-being); know what the peculiar characteristics of the soil are in the environment in which you live or work, and also have information on the technical and scientific tools that can be put in place to safeguard it in quantity and quality. In addition to air and water, natural capital is also made up of soil and its specific characteristics and qualities. These specific characteristics are the basis of a series of processes - cycle of nutritional elements, water cycle, biological activity, structure formation, gas exchange - through which the soil is able to carry out very important functions such as the regulation of the microclimate, the sequestration of carbon, the creation of a water reservoir, the supply of raw materials, food and fibre, and thus contribute to providing "ecosystem services"[10]. The functions that the soil performs, and the connected ES, vary in space, in relation to the characteristics of the soil, and in time, in relation to the conditions (climatic, management, etc.) of the context. In fact, different soils provide different services and/or of different quality. In an urban area, the most important ES are the regulation ones which recall the theme of sustainability and urban metabolism: the energy that is managed within the ecosystem

[9] https://www.snpambiente.it/snpa/consumo-di-suolo-dinamiche-territoriali-E-servizi-ecosistemici-edizione-2023/.

[10] Ecosystem services are divided into 4 categories: 1) support life, because they host plants, animals and human activities; 2) supply, because they produce biomass and raw materials; 3) regulation of hydrological and bio-geochemical cycles; 4) cultural values, as a historical-archaeological archive and a fundamental part of the landscape.

engine is returned to a level that can be used by the engine of the nearby ecosystem. The notion of ES is closely linked to the concepts of sustainable planning, green strategies, and Nature-Based Solutions (NBS). Indeed, economic prosperity and well-being strictly depend on the state of the natural resources that surround us, the so-called natural capital, and on the ecosystems that provide essential goods and services.

To preserve natural resources and ES, the help of green infrastructures is used. Urban green space is a component of green infrastructure and consists of a service that cities provide to citizens to promote their well-being. The most common definition of urban green space was given by the *European Urban Atlas* which considers urban green areas as public green areas used mainly for recreational purposes, such as gardens, parks, suburban natural areas and forests, or green areas bordering urban areas managed and used for recreational purposes.

According to the definition given by the European Union, green infrastructure is a strategically planned network of natural and semi-natural areas with other environmental elements, designed and managed to provide a wide range of ES.

Among the different types of urban green spaces, it is possible to recognize a variety of natural spaces: large and small, public, and private, simple, and complex, which when combined form a green network (for example, natural open spaces, river areas, forests, parks, gardens, squares, vegetable gardens, rows of trees, urban greenery, ponds, green roofs and green walls).

This includes green spaces and other physical elements present on land or in the sea.

Among the green infrastructures on land, we have rural and urban ones which include the Natura 2000 network, parks, gardens, hedges, vegetated strips, artificial elements such as hanging gardens, green walls, ecological bridges[11]. From an ecological-environmental point of view, green infrastructures reduce the fragmentation of natural habitats, increase the degree of biological diversity, increase self-regenerative capacities, reduce the ecological footprint of cities, mitigate the effects of climate change, reduce the effect of heat island present in the city. Furthermore, they support the spread of cycle and pedestrian mobility, fuel short-chain agriculture, improve liveability and recreational activities.

In general, it is possible to categorize the benefits produced by green areas as *environmental*: improvement of air pollution and the urban heat island effect; *social*: natural features can play an important role in residents' sense of belonging to the community and through interaction; *health*: individuals living in areas with a shortage of green spaces may be more vulnerable to stress. In fact, as a positive consequence, there has been a reduction in the number of hospital admissions caused by cardio-respiratory diseases; physical: one of the main determinants of physical activity is access to green spaces.

Today, the main reference that establishes the principles of sustainable development, the 2030 Agenda, with its 17 Sustainable Development Goals (SDGs), establishes common objectives that the member states of the United Nations are committed to achieving. Objective 11 *Sustainable cities and communities* is specifically designed to describe the

[11] Report from the Commission to the European Parliament, the Council, the European Economic and Social Committee and the Committee of the Regions - Review of progress made in implementing the EU green infrastructure strategy - Brussels, 24.5.2019 COM.

strategies to be implemented to make cities and human settlements inclusive, safe, long-lasting, and sustainable. However, the implementation of interventions that contribute to the construction of *urban green infrastructures* can bring numerous positive effects, which go beyond Objective 11 alone, therefore contributing to the achievement of other objectives among those envisaged by the 2030 Agenda, as they are capable of ensure: *environmental benefits*, provision of clean water, removal of pollutants from air and water, improved pollination, protection against soil erosion, increased pest control, improved soil quality; *benefits of climate change adaptation and mitigation*, flood mitigation, strengthening ecosystem resilience, carbon storage and sequestration, mitigation of the effects of urban heat islands, prevention of disasters due to extreme events; *biodiversity benefits*, improved wildlife habitats, ecological corridors, landscape permeability; *social benefits*, better human health and well-being, job creation, diversification of the local economy, more attractive and more liveable cities, integrated transport and energy solutions, better tourism and leisure opportunities.

From the recent debate on ES it emerges that they have great potential as an interpretative tool at the service of planning; their functions make clear society's dependence on natural resources and their substantial irreplaceability. Furthermore, the debate on ES focuses on the recognition of the socio-ecological value of natural components; on the involvement of a plurality of subjects; on the need for a holistic interpretation of the man-nature relationship. ES constitute a central theme for the development of good planning practices aimed at introducing regenerative solutions to urban systems.

To plan first and act later, it is therefore useful to know the characteristics of the soil of a specific territory and the relative quantity and quality of ES. A paradigm shift is therefore necessary. Territorial planning must protect the soil and its ES; it must also promote the recovery and restoration of those soil functions which in the urban context can help authorities to implement innovative strategies for integrating ES into urban development and regeneration in response to climate adaptation. Starting from the soil, which is the non-renewable resource par excellence, therefore, the most effective actions on the adaptive capacity of urban systems concern the management of natural capital, and require tools and solutions that use urban greenery, soil permeability, rainwater management and sustainable mobility.

But how can this change be counteracted? Contemporary urban planning is urged to search for new paradigms more suited to the scenario of malaise in which we live. The first strategic axis concerns "mitigation", measures aimed at reducing greenhouse gas emissions, a type of intervention that concerns the production, mobility, energy and land use sectors, i.e. those most responsible for the increase of emissions. The second axis concerns "adaptation", a strategy to be implemented in a preventive or reactive way, to reduce the extent of damage.

2 Integration of Knowledge for the Spatial Analysis of Functions

The *evaluation* and *mapping* of Ecosystem Services are scientific methodologies that offer an approach to the prevention of ecological problems caused by human action and the resolution of conflicts resulting from changes in land use. The approaches for the spatial analysis of the functions of these services use specific indicators. These are no different from other environmental indicators, but they focus on the environment in a different way: in addition to biophysical data, they try to capture how this information can be interpreted with respect to the benefits of nature for humans. This implies that ES indicators are often composite indicators, i.e. they combine various measurements of the supply and use of a benefit provided by an ecosystem.

The attempt to integrate the concept of ES into policies and decision-making processes has guided scientific research in the identification of increasingly developed and complex methods to evaluate and monitor these services in space and time with the aid of technological innovations, to example in the field of remote sensing and big data. The evaluation of ES is carried out by means of quantitative indicators, which constitute a simplification of complex biophysical, social and economic phenomena[12]. To map an indicator, it is necessary to acquire a series of information, which must be subjected to a *harmonization process,* in terms of scale and precision. The same indicator can vary significantly depending on the seasonality and the unit of representation of the information. The operational methodologies using software allow you to analyze large quantities of data easily and quickly. The methodology for processing the data concerns the ability to *produce spatialization of biophysical values* which makes it possible to make explicit the ability of a soil to perform or not perform ecosystem functions in such a way that all this intercepts the urban planning project. ES are synergistic, each one can help the other to make the system work and it is therefore important to understand the needs and functions of a specific urban and territorial structure.

From the 1970s onwards, the land has been occupied, without understanding how it works, most suited to the balance of the system. Here, then, this approach serves to reverse this trend, that is, to ensure that the natural capital needed to have a good quality of life (air, water, soil, etc.), is not used improperly within of the urban ecosystem.

To date, there are a range of methods and technologies (integrated modeling, open data, etc.) that demonstrate the momentum of research in attempting to quantify and monitor the natural capital and ES from which society benefits, with the aim of improving understanding of socio-ecological systems. Progress has been made with the development of tools (web platforms) that use artificial intelligence systems or that employ technologies, such as *machine reasoning and machine learning,* to model problems and phenomena of a socio-economic and environmental nature, such as capital mapping natural and ES. These platforms use an approach in which data and models are annotated as concepts and processed by artificial intelligence, to produce new knowledge and encourage its sharing and dissemination.

To obtain a more accurate quantification of the ES supply, models and software are often used, based on changes in land use and land cover, such as ARIES (ARtificial Intelligence for Ecosystem Services), InVEST (Integrated Valuation of Environmental Services and Tradeoffs) and SOLVES (Social Values for Ecosystem Services). ARIES is an open-source tool that allows you to encode ecological and socioeconomic data to map the supply, use and flows of ES using geographic information systems (GIS). The INVEST software includes fifteen models that analyze different processes that occur in terrestrial and marine ecosystems, and which are the basis for the estimation of ES, while SOLVES is a functional tool for evaluating, mapping and quantifying the perceived

[12] B. Burkhard, & J. Maes, 2017; B. Burkhard, M. Kandziora, Y. Hou, & F. Müller (2014).

social values of ES, such as aesthetics and recreation. I-Tree is another famous tool for quantifying the benefits and values produced by trees and can also be used at smaller scales (roads, plots, etc.). I-Tree provides several tools to evaluate specific benefits at different scales[13].

In a broader sense, scientific research and related technological innovations are called to guide the transition of growth and development models traditionally focused on economic performance, towards an integrated approach between ecology and economy, oriented towards the well-being of individuals and social sustainability, economic and environmental.

From a planning point of view, there are practices, which are increasingly consolidated, for strengthening resilience and adaptation objectives. There is now a wide range of technical solutions and implemented interventions that are part of NBS, which can carry out decisive tasks to increase resilience, and of establishing "adaptation solutions" on the territory through urban infrastructures. It is necessary, therefore, to plan and design for *objectives and requirements*: the ES are not new urban planning standards, but they are a new approach, a new paradigm capable of addressing the planning of cities for objectives, having in mind the awareness of which requirements can and should be asked of urban infrastructures. From an operational point of view, this means integrating traditional urban planning regulations with settlement structure rules that highlight areas of the city equipped with ecological-environmental standards in such a way as to pursue, for each urban regeneration intervention, the maintenance and valorisation of those indispensable ES in the urban context. It is therefore necessary to make this type of ES approach effective by integrating it into the vast phenomenology of data, information and variables that distinguish the process analysis-evaluation-project in planning activity.

Having recognized the emergence of a new 'urban question', the topic of soil today poses new and broader questions aimed at limiting its consumption and redefining its design contents. It is therefore on the awareness of having to harmonize the maintenance - or reconstitution - of *natural capital* and the compatibility of the *fixed capital* of settlement and infrastructural systems within new practices, that the growing interest and commitment towards the theme of ES is contextualized. Qualitative knowledge related to ES appears essential for the preliminary assessment of environmental effects- and the consequent economic and social impacts - which derive from the possible land use arrangements. By estimating the ecosystem value differentials both in quantitative and qualitative terms, managing to evaluate the quality rather than just the quantity of transformed soil: an evaluation that offers useful knowledge to support choices, policies and projects for ecologically oriented, urban and territorial regeneration support of urban planning choices that take into account the containment of land consumption and the protection of its ecosystem functions.

[13] Marino D., Poli D., Rovai M. (cura di) 2022.

3 Concluding Considerations

In conclusion, it is possible to formulate some general considerations useful for summarizing the concepts presented so far. It is necessary for all the actors involved in the planning process to understand in a profound and coherent way the critical issues and possibilities they are facing. The containment of land consumption, the difficult reproducibility of some environmental units and their respective ES, the need to operate environmental mitigation and compensation where a transformation involves a worsening of environmental quality, constitute some of the numerous pieces that make up the mosaic of sustainable urban planning.

Cities as complex ecological systems constitute the main place of exploitation of ES and at the same time the privileged environment from which to start again to implement resource management that is proactive and directed by targeted policies.

These reflections underline, therefore, the importance of incorporating multiple perspectives in the evaluation of ES, capable of considering the different dimensions of value in the study of ES to support *decision makers* in defining actions aimed at increasing the quality of life in cities. Biophysical, ecological, and economic values have been widely studied, but there are critical issues in the evaluation of intangible values, such as sociocultural ones, and in the integrated consideration of all dimensions of value. Indeed, a pluralistic vision of ES research has become more important than ever, especially when it becomes a tool to support the planning and transformation of cities in a sustainable way. The proposed methodological approach considers not only the biophysical and ecological value provided by ES, but also their translation into economic terms, integrated with values of a sociocultural nature.

Ackowledgements. This research work is the result of the research activities conducted within the following PNRR research project funded by the European Union – NextGenerationEU: "Pilot Project 4.6.1 "Climate adaptation plans for the reduction of the ecological footprint and ecological debt, aimed at improving the conservation and transformation response in terms of resilience and quality of life in the Calabrian and Lucanian urban and territorial systems" (Goal 4.6 - Planning for Climate Change to boost cultural and natural heritage: demand-oriented ecosystem services based on enabling ICT and AI technologies - Tech4You Innovation Ecosystem), Finanziato dall'Unione europea- Next Generation EU, Missione 4 Componente 2 - CUP C33C22000290006 (ECS_00000009).

References

1. Nebbia G.: L'ecosistema urbano. Economia e Ambiente, 1–2/2006
2. Nicoletti M.: L'ecosistema urbano, Dedalo, Bari 1978, pp. 14
3. Crutzen, P.J.: Human Impact on climate has made this: the "Anthropocene Age". New Perspect. Quaterly **22**, 14 (2005)
4. Giaimo, C.: Performance Urbane e Servizi Ecosistemici. Verso nuovi standard? In Dopo 50 anni di standard urbanistici in Italia. Publisher INU Edizioni 2019
5. Giaimo, C., e Salata, S.: Rigenerazione urbana e buon uso del suolo: mappare e valutare i servizi ecosistemici alla scala locale. Esperienze dal Progetto Life Sam4cp, in urban@it (2017)

6. Munafò M. a cura di (2020), Consumo di suolo, dinamiche territoriali e servizi ecosistemici. Report SNPA, 15/20
7. Burkhard, B., Maes, J.: Mapping ecosystem services, Sofia. Pensoft (2017). https://doi.org/10.3897/ab.e12837
8. Burkhard, B., Kandziora, M., Hou, Y., Müller, F.: Ecosystem service potentials, flows and demands-concepts for spatial localisation, indication and quantification. Landscape Online (2014)
9. Marino, D., Poli, D., Rovai, M.: (cura di), Montagna, servizi ecosistemici e strumenti di governance in Toscana, Ecosistemi-Politica ambientale [della] Regione Toscana, Università di Firenze (2022)
10. Dell'Ovo, M., Corsi, S.: Urban ecosystem services to support the design process in urban environment. a case study of the municipality of milan. Aestimum **2020**, 219–239 (2020)
11. European Environmental Agency (EEA), Nature-based solutions in Europe: Policy, knowledge and practice for climate change adaptation and disaster risk reduction. Copenhagen, Denmark (2021)
12. Caprioli, C.: Urban ecosystem services: a review of definitions and classifications for the identification of future research perspectives. In: 20th International Conference on Computational Science and Ap- plications (ICCSA 2020), July 1 – 4 (2020)

Implementing the Ecological Transition: Green Infrastructure and Ecosystem Services as Key Factors for Urban Planning Innovative Paradigms

Gabriella Pultrone(✉)

Mediterranea University of Reggio Calabria, Via Melissari 25, 89124 Reggio Calabria, Italy
gabriella.pultrone@unirc.it

Abstract. The paper proposes some reflections resulting from the author's ongoing research activities on Green Infrastructure and Ecosystem Services, which, when integrated into urban planning, can improve the supply of goods and services for the well-being of society, by incorporating the ability to adapt to current risks and future climate change, reducing the ecological footprint and ecological debts, while improving resilience, health, and quality of life, and leaving no one behind, in line with the UN 2030 Agenda. Despite the relevance of the topic and the critical feature of the "time factor" in addressing a climate crisis requiring urgent responses and far-sighted policy choices, and the numerous international documents and commitments, scientific advances have not yet produced adequate operational impacts, which are fully integrated and widespread in policies, strategies, and actions at a local level, in order to move towards the necessary ecological transition in the perspective of integral ecology. Nevertheless, in the EU scenario, there are many interesting experiences aimed at an ecological transition that goes beyond the technological and energy transition and concerns foremost the climate issue and the exit from the fossil fuel system, but also the dramatic loss of biodiversity, conflicts over access and use of resources, the international geopolitical instability, and profound economic and social inequalities. In this ongoing process, cities are drivers of innovation and, through the selected case studies proposed later in this paper, it is possible to highlight the lessons learned in terms of critical and/or success factors; identify some guidelines to adapt possible solutions to the specific issues to be addressed in different contexts; suggest future research and in-depth perspectives.

Keywords: Ecosystem Services · Ecological Transition · Green Infrastructure

1 Background and Challenges to Guide the Green Transition

According to the United Nations, the urban population is expected to increase by almost 600 million by 2030, reaching a total of 5.2 billion and representing 60% of the world's population (https://population.un.org/wup/). Urban areas occupy less than 1% of the Earth's surface but are home to more than half of its population. They are ecosystems

F. Moraci et al. (Eds.): NMP 2024, LNNS 1272, pp. 138–150, 2025.
https://doi.org/10.1007/978-3-031-82927-7_9

representing a radical transformation of the natural areas, whose condition profoundly positively or negatively affects the quality of life. As a matter of fact, on the one hand, functioning ecosystems help to clean the air and water, to cool urban heat islands and to support our well-being; they offer opportunities for rest and play, and can host a surprising amount of biodiversity. On the other hand, they can be highly degraded due to bad planning, urban sprawl that devours natural habitats and fertile agricultural land, sealing soils leaving little space for green infrastructure, waste and emissions from industry, traffic and homes that pollute waterways, soil, and air. Therefore, the restoration of urban ecosystems requires awareness and commitment on the part of citizens and decision-makers, who should increasingly place green and blue infrastructure (GBI) at the centre of urban planning, in accordance with international commitments, whose implementation a local commitment, which makes cities and territories protagonists and promoters of the green revolution and ecological transition.

This is why the concept of Green Infrastructure (GI) has the potential to promote planning and implementation of multifunctional green and blue spaces that tackle several urban sustainability issues. GI is widely understood as a network of physical features providing ecological, economic, and social benefits to society through Nature-Based Solutions (NBSs). GI is closely linked to the quality of the soil, in which a multitude of organisms operate providing us with food, biomass, fibres and raw materials; regulate the cycles of water, carbon and nutrients; and make life possible on earth [1, 2]. Soil is a vital, limited, non-renewable and irreplaceable resource, as well as an essential basis of the economy, society and the environment, as it produces food; increases our resilience to climate change, extreme weather events, droughts and floods; supports our well-being; stores carbon; has a greater capacity to absorb, conserve and filter water; and provides vital services, such as safe and nutritious food and biomass for the non-food sectors of the bioeconomy [3, 4]. In this context, urban and territorial regeneration is often proposed as a useful tool for avoiding land consumption, and certainly represents a priority for rethinking, in the direction of an increasingly indispensable and urgent ecological transition, the order of a more and more fragile world poorly equipped to face the great challenges posed by climate change and the widespread environmental and landscape degradation. In order to be sustainable, regeneration should intervene on what already exists, without causing the loss of portions of soil that are not yet artificial and should be considered as limited and non-renewable resources, as precious assets for the community, to be enhanced in their natural functions also through the above-mentioned regeneration initiatives.

Among the most significant impacts linked to the increasing urbanisation and the densification of urban areas, the reduction in the availability of public green spaces takes on a central role in the definition of liveable cities, not only in environmental terms (*i.e.*, contributing to biodiversity, to the improvement of air quality, to climate mitigation, and to water infiltration), but also in terms of social cohesion, interaction, equity, and equality. Hence, the theme of the implementation of public green areas, especially in expanding urban contexts, has been addressed both at a global and national level through the adoption of numerous and various thematic strategies, also linked to the UN 2030 Agenda and the *One Health* approach (https://www.who.int/news-room/questions-and-answers/item/one-health).

Thanks also to the *European Strategy for Biodiversity towards 2020*, Ecosystem services (ES) are increasingly at the centre of innovative planning practices that take into due consideration the benefits deriving from the natural component. Many studies have analysed the relationships between the provision of ES and the characteristics of green areas and of settlements in urban contexts, where those identified as the most relevant were heat mitigation, carbon capture and storage and runoff control, with particular reference to flood-related events. In general, the results show that there is ample space to improve the performance of ecosystem services based on fine-tuning measures involving building and population density and vegetation cover [5]. In this direction, the *Nature Restoration Law*, approved by the European Council last November 9, 2023, is one of the four pillars of the European 2030 biodiversity strategy. It promotes biodiversity and climate action across Europe, beyond the introduction of regulatory tools for the prevention and protection of human health, and the integration of the environmental policies in force. As part of the *European Green Deal*, the objective is the restoration of ecosystems and the strengthening of biodiversity to ensure the well-being of current and future generations. Thus, it is possible to create advantages for both biodiversity and the rebalancing of the territory, to fight the climate crisis and to pursue a more effective conservation of the natural capital and the ecosystem services it produces, favouring adaptation and mitigation policies. Therefore, it provides an important reference for urban planning, considering the provisions about the absolute limitation of any loss of green areas within cities, as well as the other objectives aimed at directing and supporting policies for urban renaturation and forestation. GI become a central element in the redefinition of urban planning strategies, through actions to restore and enhance ecosystem services, with an approach increasingly oriented towards a qualitative and performance evaluation of the ecological functions of urban soils, together with the use of NBSs, which reconfigure the urban planning project in an ecological sense.

The mapping and biophysical and economic evaluation of ES is the starting point for the definition of territorial governance policies, which, at the territorial scale, are aimed at the protection of agroforestry ecosystems, while, at the local scale, are aimed at the containment and good use of the soil resource. Thus, the assessment of the effects of the different choices becomes essential and involves the estimation of the costs and benefits associated with different land use scenarios and/or protection policies and guidelines specific to the territorial and urban planning tools, trying to integrate the programmatic framework and regulations of higher administrative levels and to implement synergistic urban planning actions at different scales.

In the light of the above, the paper proposes some reflections resulting from the author's ongoing research activities on green infrastructure and ecosystem services which, when integrated into urban planning and design, can improve the supply of goods and services for the well-being of society by incorporating the ability to adapt to both current risks and future climate change, reducing the ecological footprint and ecological debts, while improving resilience, health, and quality of life [6–9]. Despite the relevance of the topic and the critical feature of the "time factor" in addressing a climate crisis, which requires urgent responses and far-sighted policy choices, the numerous international documents and commitments, and scientific advances have not yet produced adequate operational impacts, fully integrated and widespread in policies, strategies, and actions at local level, towards the necessary ecological transition in the perspective of integral ecology.

Nevertheless, there are many interesting experiences aimed at an ecological transition that goes beyond the technological and energy transition and concerns foremost the climate issue, as shown in the second paragraph.

2 EU Cities as Players of the Transition: Work in Progress

As mentioned above, cities must tackle global challenges, above all those linked to climate, pollution, and energy consumption, in order to increase the capacity of the urban environment to deal with the effects of these alterations, which, more or less directly, also manifest themselves in terms of socio-spatial inequity and injustice in the distribution and access to resources and services. According to the ESPON Policy Brief "Green infrastructure in urban areas" [10], between 2006 and 2012, many cities throughout Europe lost green spaces, mainly because of unsustainable urbanisation. At the same time, however, they have the means to counteract the degradation and loss of natural capital and the ecosystem services provided by this capital. As a matter of fact, local authorities, responsible for planning and investing in urban infrastructure, can make important choices concerning the nature of infrastructure construction, renewal, or expansion, and can promote greener, more sustainable cities. A GI-based approach to spatial planning is needed, which not only connects different elements of the nature, but also goes beyond ecological and political boundaries for connecting sectoral policies. Therefore, this section is focused on some significant case studies to understand how GI, ES and NBS are used and what is still needed to improve planning decisions. In fact, case studies in the field of planning show that plans, programmes or policies can trigger virtuous transformations, or bring with them risks when concretely implemented (Campbell, 2003). Naturally, the analysis is not exhaustive as it shows a selected part of ongoing research to be further developed. Yet, the first conclusive considerations already outline both the lessons learned, in terms of critical and/or success factors, and the possible future research perspectives. In this regard, the ESPON GRETA project (*GReen infrastructure: Enhancing biodiversity and ecosysTem services for territoriAl development*) is of particular interest, since it has implemented a territorial model of potential green infrastructure in European cities, carrying out an assessment of the green infrastructure in urban spaces, including all available green and blue areas. The most relevant land cover/use dataset for this analysis of cities and their immediate hinterland (peri-urban space) is the Urban Atlas layer provided by the European Copernicus programme. The data complements city statistics collected by Eurostat in the framework of the Urban Audit Programme, where cities are represented on three spatial levels: the *core city*, a local administrative unit (LAU), in which the majority of the population lives in an urban centre of at least 50,000 inhabitants; the *functional urban area* (FUA), which adds the commuter area to the city; the *greater city*, which approximates the urban centre when this stretches far beyond the administrative city boundaries. To provide an overview of the state of urban green infrastructure, the following parameters and indicators were calculated and mapped: share of green urban areas (GUA) within the core city (which represents the level of the city); the entire FUA (which represents the entire reference unit); the FUA without the core city (which represents only the peri-urban space; all values are expressed in %); and the ratio between the share of GUA within the core city and the share of GUA within the entire FUA (ratio without units).The project aims to support European, national, regional and urban authorities to better understand how to create, manage and improve green infrastructure in urban areas using an integrated approach. The need arises from the fact that, despite the numerous good practices of spatial planning systems that incorporate a GI approach, there are still uncertainties in practice regarding the scales and phases of the process.

The analysis of GI policies and planning in Europe leads to the definition of good practice as the implementation of tools and/or actions in a way that allows us to perceive the optimal increase in connectivity and multifunctionality in green spaces. The 25 case studies identified allow describing the ways in which modern tools, policies, and processes for implementing GI development are used by the local or regional authorities, and the ways in which local and regional authorities interact with private sector actors and local stakeholders for the successful implementation and management of GI. Table 1 sums up 11 significant selected examples, from which the identification of some key policy messages arises.

A first analysis of possible innovative models of governance and territorial planning of GI in European cities shows that the most effective actions on the adaptive capacity of urban systems concern the management of natural capital, especially in the context of more or less densely built spaces (public and private), and require tools and solutions that use urban greenery, soil permeability, the management of theoretical waters and also sustainable mobility [12]. This is the case of Lisbon (see ID 1 in Table1), which has also applied the city's biodiversity index; took part in the MAES urban pilot project (mapping and evaluation of ecosystem services) almost ten years ago; and played the role of citizen laboratory in the EnRoute project of the EU Joint Research Centre.

Furthermore, the other ten examples in Table 1 refer to integrated, multi-scale and multi-sectoral activities through the articulation of planning tools (Basque Country ID 10); inter-institutional collaboration in spatial planning (Finnish cities-ID 2, in Poznań-ID 3, London-ID 7); multi-actor collaborative approach involving companies and/or associations (Bologna-ID 4, Amsterdam-ID 8, Ljubljana-ID 6); incentives for families (Bratislava-ID 5); and implementation of green infrastructure solutions in built-up environments (Malmö -ID 9) and coastal cities (The Netherland-ID 11).

The attention paid to these aspects derives from the fact that, with reference to the main outcomes of the project, the first three categories of the thirteen most frequent barriers and challenges in the implementation of GI are the following: lack of a strategic vision, of common objectives and integrated planning; lack of long-term experience; and financial incentives. In general, the following key policy messages are highlighted to identify areas where measures are needed to protect green spaces from disappearing and to implement green (and blue) infrastructure: continuous monitoring of GI development; a strategic vision widely shared by policymakers and planners and implemented through an integrated planning process embedding different sector policies and different levels of governance. An integrated approach must establish connections between different natural elements in the geophysical area, between nature and people's quality of life, across ecological and political boundaries, and between policy sectors, given the growing

Table 1. Selected examples of integration of urban GI into policies, plans, and projects. Author's elaboration from ESPON GRETA *Green Infrastructure: Enhancing Biodiversity and Ecosystem Services for Territorial Development* [11]

ID	Country	Case Study/Best Practice/ Project Title	Main aspects/Goals
1	Portugal	Urban strategy for biodiversity in Lisbon	Counteract habitat fragmentation and climate change by increasing soil permeability and mitigating the "heat island" effect
2	Finland	Integrating conservation of urban GI into spatial planning policy in Finnish cities	Integrate urban GI conservation into national planning policy in a coherent way. The Finnish Ministry of the Environment coordinates the process of developing national urban parks, established to preserve the beauty of the cultural landscape, to maintain ecological corridors, biodiversity and cultural and natural heritage in urban areas
3	Poland	Pooling of public funding to introduce NBSs in Poznań	Innovative collaboration between the City Hall, the Project Coordination, the Urban Regeneration Office and the Department of Education to introduce NBSs in the gardens of state-run preschools in the densely populated city centre area and encourage them to remove hard surface seals, introduce more biodiversity and create natural gardens linked to other green urban corridors
4	Italy	Carbon footprint compensation scheme to finance tree planting in Bologna	Through the Green Areas Inner city Agreement (GAIA), an outcome of a LIFE project, companies can calculate their carbon footprint and compensate for it (voluntarily) by making donations towards tree planting

(*continued*)

Table 1. (*continued*)

ID	Country	Case Study/Best Practice/ Project Title	Main aspects/Goals
5	Slovakia	Scheme for the purchase of rainwater management installations in Bratislava	As part of the *Bratislava Turn Green* project (Bratislava goes green), the Municipality encourages families to contribute to protecting the city from pluvial flooding through a subsidy scheme for the purchase of stormwater management systems. Most of successful applicants have installed rainwater catchment tanks, created rain gardens, replaced impermeable surfaces with permeable materials or installed green roofs
6	Slovenia	"Beyond a construction site" - community-based gardening in Ljubljana	In 2010, the cultural association *Obrat*, in collaboration with the cultural organization *Bunker*, approached the municipality (owner of the site) to gain temporary lease of land consisting of a derelict construction site and transform it, with the help of residents, into an attractive community space intended for urban gardens, socialization, education, and culture
7	United Kingdom	Integrated planning for GI in London	The *Olympic Park* in London, intended to host the Olympic Games and Paralympics in 2012, provides a valuable and high-profile example of how GI is designed in the Lower Lea Valley brownfield site, within a dense and complex urban area, thanks to a framework clear strategy and high-level political commitment, which acknowledges the importance of establishing specific targets for biodiversity at the start of the process

(*continued*)

Table 1. (*continued*)

ID	Country	Case Study/Best Practice/ Project Title	Main aspects/Goals
8	The Netherland	Collaborative multi-actor approach to greening the city of Amsterdam	Creation of the *Amsterdam Rainproof platform* by the water supply company *Waternet,* in collaboration with citizens, public employees, and entrepreneurs. It raises awareness among urban residents and policy makers of the consequences of soil sealing and contributes to addressing the challenge of flooding caused by stormwater, encouraging people to consider the possibility of extreme precipitation when designing homes, gardens, streets, and parks. A group of social entrepreneurs, *Roof Doctors,* introduced the concept of *polder roofs,* which transform unused roofs into places intended for the development of nature, recreation, water storage and food and energy production
9	Sweden	Green space factor for implementing green and blue infrastructure in built-up environments in Malmö	Malmö municipal authorities have developed a formula for developing greener housing blocks, including the green space factor in many projects and as part of the local authority's environmental building practice. Developers need to compensate every surface they want to seal with something else that is green or blue. The tool measures the way ecosystem services are produced and emphasizes the assessment of the reduction of noise and air pollution and water purification

(*continued*)

Table 1. (*continued*)

ID	Country	Case Study/Best Practice/ Project Title	Main aspects/Goals
10	Spain	GI considerations for climate change adaptation in regional spatial planning guidelines, Basque Country	The Basque Country has a robust spatial and urban planning system that is integrated, multi-scale and multi-sectoral thanks to the various existing planning instruments; it uses operative governance mechanisms; it includes complementary competence distribution between public administrations (regional, provincial, local); it is characterised by a territorial management culture; it includes strong consideration of natural capital protection and GI enhancement, with substantial activities in the field of NBSs and key complementary activities, *e.g.*, consideration of health in urban design
11	The Netherlands	Soft engineering for coastal management in the Netherlands	The annual Dutch Delta program aims to ensure the sustainability and robustness of flood risk management and freshwater supplies beyond 2050, setting new flood protection and territorial adaptation standards. In recent years, the dike system used to protect land has provided the opportunity to implement NBSs, together with smart land management. The Delta program has demonstrated that sustainable coastal soft engineering solutions for coastal zone protection include a combination of beach nourishment, dune replenishment and the planting of vegetation to stabilize the newly replenished beach and dune

interrelationships between places and the need for proactive and strategic planning that goes beyond administrative boundaries. In light of the above, Strategic Environmental Assessment (SEA) is considered as a policy tool to incorporate green infrastructure into strategies, plans and programmes, which could help to establish a common framework across EU states, strengthen strategic thinking and the centrality of ecosystem services, thus increasing territorial resilience.

Equally fundamental is to take into account the synergies and trade-offs between ecosystem service and to be aware of these relationships in order to prioritize their effects based on the best available knowledge. As a matter of fact, ecosystem services often appear in packages that, under certain circumstances, reinforce each other (*i.e.*, they are in synergy with each other), while, in other cases, they can influence each other negatively (*i.e.*, there are compromises between them).

Another aspect of undoubted importance, which cannot be addressed here but is worth of further specific research, concerns the cost-effectiveness ratio and the feasibility of the various interventions, in relation to their scale of governance.

In a nutshell, adequate methods to assess urban ecosystem services at the right scale and resolution while also accounting for the multi-functionality of urban green infrastructure are still needed. On the other hand, future urban plans would benefit from a further appropriation of the ecosystem service approach by practitioners and decision-makers [13–16]. Acknowledging the whole range of urban ecosystem services, defining strategic objectives related to their provision, and explicitly identifying demand and beneficiaries could increase awareness of the values at stake, ensure long-term commitment in the implementation phase, and strengthen planning arguments against any conflicting interests.

3 Discussion and Conclusions. Beyond Transition

From the path outlined so far, the interdisciplinary nature of the issues relating to GI and SE, and their potential for planning purposes in the ongoing ecological transition process has emerged as a fertile ground for future research. The World Cities Report 2022 "Envisaging the Future of Cities" [17] aims to provide greater clarity and insights into the future of cities based on existing trends, challenges, and opportunities, as well as disruptive conditions, and suggests ways in which cities can be resilient, because they are better prepared to face a wide range of shocks and transition towards sustainable urban futures. That is why, informed preparation is essential, which offers the opportunity to anticipate change, correct the line of action, and become more aware of the different scenarios or possibilities that the future of cities offers, in a vision that should embody the "new social contract" in the form of universal basic income, universal health coverage, universal housing and basic services for all.

Another aspect worthy of attention is that, if, on the one hand, the "green transition" aims to achieve sustainability and tackle climate change and environmental degradation, on the other, the growing importance of digital technologies is transforming societies and economies. Moreover, through the "digital transition", technologies should be oriented towards sustainability, prosperity and the empowerment of citizens and businesses. Successfully managing the twin transitions, green and digital, is the cornerstone for offering a sustainable environment and a more equitable and competitive future but, to unlock their potential and prevent negative effects, they must be achieved together with proactive and integrated management [18].

In a broader sense, the biggest supporting factor in GI implementation is a strategic vision that is widely shared by policymakers and planners and is implemented through an integrated planning process that embeds different sector policies and different levels of governance [9]. It is clear that GI should not be promoted as a sectoral element, but with the awareness that it contributes also to improve strategic thinking, ecological processes and their benefits, positioning itself as a planning criterion relevant to a more resilient territorial development. Spatial planning is, by its very nature, an intersectoral discipline, which includes various thematic issues to achieve balanced territorial development. Therefore, it presents an optimal predisposition to integrate the planning, implementation, and management of GI. There are many possible ways and tools by which local authorities can facilitate the implementation of GI by private stakeholders like infrastructure builders and homeowners through regulatory and planning tools. For example, zoning regulations can mandate that new residential neighbourhoods include a certain percentage of green space, or existing regulatory requirements can be leveraged to allow investment in NBSs rather than grey solutions. Equally interesting is the robust climate strategy approved by the Basque Country, which includes explicit actions to integrate adaptation into territorial planning and to deploy resources towards resilient urban development. In particular, the development of substantive information on climate risks and impacts (climate projections, flood risk maps, urban heat island studies, local vulnerabilities) has been crucial to enable cutting-edge operational consideration of climate adaptation in territorial and urban planning tools. Another fertile input for theoretical reflection and applied research is the relationship between ES and urban planning standards, as an integrated and multifunctional approach to standards could provide answers to multiple issues, such as hydraulic risk, air, water and soil pollution, loss of biodiversity. In a nutshell, the future research on the topics of GI, ES and NBS, could aim to:

- Create place-specific planning and implementation strategies that contribute to sustainable landscape development through the simultaneous achievement of social objectives, maintenance of biodiversity and provision of important ES
- Explore how the ES concept could be used to facilitate knowledge co-production and cooperation between different actors
- Learn from planning experiments and empirical applications of case studies and make recommendations on how ES could be applied in planning tools to improve knowledge, facilitate cooperation and influence decision-making.

Finally, it should be reminded that ecological transition is an "umbrella" concept, an expression of a new lexicon, essentially intended as a technological change, that is, a set of actions for a sustainable economy to encourage the transition from a system based on non-renewable energy sources (fossil capitalism) to a virtuous model centred on the use of green sources (only green economy, green capitalism) [19]. However, to address the ongoing emergencies, a radical and deeper change in social and economic behaviour is needed to lead to the collective and individual "conversion" to new lifestyles, to the recognition that our "common home" must be protected and cared for [20]. Therefore, beyond a "transition", which contemplates the dominance of technology, and risks being only apparent, a "conversion" is necessary, which incorporates the landscape and the protection of biodiversity as founding elements for an unprecedented balance between humanity and nature for a new less anthropocentric planet.

Acknowledgements. This research work is the result of the research activities conducted within the following PNRR research project funded by the European Union – NextGenerationEU: "Pilot Project 4.6.1 "Climate adaptation plans for the reduction of the ecological footprint and ecological debt, aimed at improving the conservation and transformation response in terms of resilience and quality of life in the Calabrian and Lucanian urban and territorial systems" (Goal 4.6 - Planning for Climate Change to boost cultural and natural heritage: demand-oriented ecosystem services based on enabling ICT and AI technologies - Tech4You Innovation Ecosystem), Finanziato dall'Unione Europea - Next Generation EU, Missione 4 Componente 2 - CUP C33C22000290006 (ECS_00000009).

References

1. European Commission: EU Soil Strategy 2030. Healthy soils for the benefit of people, food, nature and climate. COM(2021) 699 final (2021)
2. Pileri P.: L'intelligenza del suolo. Piccolo atlante per salvare dal cemento l'ecosistema più fragile, AltraEconomia, Milano (2022)
3. European Commission: Proposal for a Directive of the European Parliament and of the Council on Soil Monitoring and Resilience (Soil Monitoring Law). COM(2023) 416 final
4. Neonato, F., Tomasinelli, F., Colaninno, B.: Oro verde. Quanto vale la natura in città, Il Verde Editoriale, Milano (2019)
5. Lai, S., Zoppi, C.: Factors affecting the supply of urban regulating ecosystem services. Empirical estimates from Cagliari, Italy. TeMA – J. Land Use, Mobility Environ. (2), 7–32 (2023). https://doi.org/10.6093/1970-9870/10194
6. Pultrone, G.: The ecological challenge as an opportunity and input for innovative strategies of integrated planning. In: Della Spina L Calabrò F Bevilacqua C. New Metropolitan Perspectives. Local Knowledge and Innovation Dynamics towards Territory Attractiveness through the Implementation of Horizon/E2020/Agenda2030, pp. 691–698. Springer (2019)
7. Pultrone, G.: Performance-based planning for sustainable cities. innovative approaches and practices in Italy. In: Bevilacqua, C., Calabrò, F., Della Spina, L. (Eds.), New Metropolitan Perspectives Knowledge Dynamics and Innovation-driven Policies Towards Urban and Regional Transition, 2, pp. 430–439. Springer Nature (2020)
8. Pultrone, G.: The city challenges and the new frontiers of urban planning. TeMA – J. Land Use Mobility Environ. **16**(1), 27–45 (2023). https://doi.org/10.6093/1970-9870/9392
9. Pultrone, G.: Through and beyond the Poli(s)crisis: guiding the eco-social transition in UE. UPLanD-J. Urban Plann. Landscape Environ. Des. **7**(1), 61–76 (2023). https://doi.org/10.6093/2531-9906/10417
10. ESPON: POLICY BRIEF. Green infrastructure in urban areas. ESPON EGTC, Luxembourg (2020). https://www.espon.eu/green-infrastructure-urban-areas. Accessed 19 Jan 2024
11. ESPON GRETA. Green Infrastructure: Enhancing Biodiversity and Ecosystem Services for Territorial Development. ESPON MSP-LSI (2019). https://www.espon.eu/green-infrastructure. Accessed 19 Jan 2024
12. Giaimo, C, Salata, S.: Rigenerazione urbana e buon uso del suolo: mappare e valutare i servizi ecosistemici alla scala locale. Esperienze dal Progetto Life Sam4cp. In: Background Papers. Rapporto sulle città 2017. Mind the gap. Il distacco tra politiche e città. Rivista online di Urban@it - 1/2017 (2017)
13. Milligan, Z.E.E.: Enabling the integration of ecosystem service-based approaches into planning organizations: municipal natural asset management. A thesis presented to the University of Waterloo in fulfilment of the thesis requirement for the degree of Master of Environmental Studies in Planning. Waterloo, Ontario, Canada (2019)

14. Zhang, S., Muñoz, R.F.: Assessing and mapping ecosystem services to support urban green infrastructure: the case of Barcelona. Spain. Cities **92**, 59–70 (2019)
15. Cortinovis, C., Geneletti, D.: Ecosystem services in urban plans: what is there, and what is still needed for better decisions. Land Use Policy, January 2018, **70**, 298–312 (2018)
16. Visintin, F., Navone, A., Niccolini F. (Eds.): La contabilizzazione e valorizzazione dei servizi ecosistemici e del capitale naturale nelle aree marine protette. Metodologie e prospettive di governance. Il caso Tavolara Punta Coda Cavallo, Franco Angeli, Milano (2021)
17. UN-Habitat: World Cities Report 2022. Envisaging the Future of Cities. UN-Habitat, Nairobi (2021). https://unhabitat.org/wcr/. Accessed 19 Jan 2024
18. Muench, S., Stoermer, E., Jensen, K., Asikainen, T., Salvi, M., Scapolo, F.: Towards a Green and Digital Future. Publications Office of the European Union, Luxembourg (2022)
19. Cavallo, R.: Le Parole della Transizione Ecologica, Un lessico per l'economia circolare. Ed. Ambiente, Milano (2021)
20. Papa Francesco: Lettera enciclica Laudato si' del Santo Padre Francesco sulla cura della casa comune (2015). https://www.vatican.va/content/francesco/it/encyclicals/documents/papa-francesco_20150524_enciclica-laudato-si.html. Accessed 19 Jan 2024

Author Index

F. Moraci et al. (Eds.): NMP 2024, LNNS 1272, p. 151, 2025.
https://doi.org/10.1007/978-3-031-82927-7

The manufacturer's authorised representative in the EU is Springer
Nature Customer Service Centre GmbH, Europaplatz 3, 69115 Heidelberg,
Germany. If you have any concerns regarding our products, please
contact ProductSafety@springernature.com

Printed and bound by CPI Group (UK) Ltd, Croydon, CR0 4YY
01/10/2026
02225977-0013